BIOLOGICAL OCEANOGRAPHIC PROCESSES

BIOLOGICAL OCEANOGRAPHIC PROCESSES

TIMOTHY R. PARSONS

and

MASAYUKI TAKAHASHI

Institute of Oceanography
University of British Columbia

PERGAMON PRESS

OXFORD · NEW YORK · TORONTO

SYDNEY · PARIS · BRAUNSCHWEIG

Pergamon Press Offices:

U.K.	Pergamon Press Ltd., Headington Hill Hall, Oxford OX3 0BW, England
U.S.A.	Pergamon Press Inc., Maxwell House, Fairview Park, Elmsford, New York 10523, U.S.A.
CANADA	Pergamon of Canada Ltd., 207 Queen's Quay West, Toronto 1, Canada
AUSTRALIA	Pergamon Press (Aust.) Pty. Ltd., 19a Boundary Street, Rushcutters Bay, N.S.W. 2011, Australia
FRANCE	Pergamon Press SARL, 24 rue des Ecoles, 75240 Paris, Cedex 05, France
WEST GERMANY	Pergamon Press GmbH, 3300 Braunschweig, Postfach 2923, Burgplatz 1, West Germany

First edition 1973

Reprinted 1975

Library of Congress Cataloging in Publication Data

Parsons, Timothy Richard, 1932–
   Biological oceanographic processes.

   Bibliography: p.
   1. Marine ecology. 2. Marine plankton.
3. Primary productivity (Biology) I. Takahashi,
Masayuki, joint author. II. Title.
QH541.5.S3137   1973   574.5'2636   73–7758
ISBN 0–08–017603–8
ISBN 0–08–017604–6 (flexicover)

Printed in Great Britain by Biddles Ltd., Guildford, Surrey

CONTENTS

ACKNOWLEDGEMENTS

ACKNOWLEDGEMENTS for the use of copyright material are given as follows: Cambridge University Press, Hutchinson Publishing Group Ltd. and the Fisheries Research Board of Canada for parts of Fig. 3; Botanical Society of America for Fig. 24; Academic Press Inc. Ltd. for Fig. 37 and Table 28; Botanical Society of Japan for Fig. 29; American Society of Limnology and Oceanography for Figs. 8, 13, 16, 18, 26, 35, 38, 44 and Table 11; Conseil International pour l'Exploration de la Mer for Figs. 7, 17 and 36; International Commission for the Northwest Atlantic Fisheries for Fig. 56; National Academy of Sciences for Fig. 12; Fisheries Research Board of Canada for Figs. 20, 21, 46 and Tables 4, 8 and 27; International Association of Geochemistry and Cosmochemistry for Fig. 30; Microforms International Marketing Corp. for Figs. 9, 10 and 14A; Oceanographical Society of Japan for Fig. 4; Hokkaido University for Figs. 42, 45 and Table 32; *Journal of Phycology* for Fig. 34; Springer-Verlag for Tables 6, 7, 15 and 18; Marine Biological Association for Table 9; Scottish Marine Biological Association for Figs. 1 and 9; Oliver & Boyd, and Otto Koeltz Antiquariat for Fig. 49 and Tables 25 and 31; *Journal of Marine Research* for Table 30; *Fishery Bulletin* for Table 19. Table 14, Figs. 2 and 14B are copyright 1970, 1963 and 1969, respectively by the American Association for the Advancement of Science. Fig. 6 and parts of Fig. 3 were originally published by the University of California Press; reprinted by permission of the Regents of the University of California. Tables 1 and 5 are published with permission from George Allen & Unwin Ltd; Fig. 15 is with permission From the Koninklijke Nederlandse Akademie van Wetenschappen; Fig. 47 is with permission From the North-Holland Publishing Co.

INTRODUCTION

THE FOLLOWING text is intended to serve as an introduction to the field of quantitative biological oceanography. The title of the book has been chosen to reflect this fact and the book should not be regarded as a literature review. We have selected reference material illustrative of certain types of biological oceanographic processes which can generally be explained either in terms of an empirical equation or through the use of definite biological or chemical descriptions. By definition these empirical relationships are advanced as being among the most acceptable at the time of writing the text but it is to be expected that researchers will improve or disprove many of the processes discussed, in the light of further scientific advancement. Such is the nature of science. As an introduction to the subject, however, we feel that students, physical oceanographers, engineers, hydrologists, fisheries experts, and scientists in a number of other professions may require some quantitative expressions of biological oceanographic phenomena. In some cases we have drawn on examples from the freshwater environment; in practically all examples, however, we have referred to processes in the near-surface pelagic environment, and benthic, littoral, coral reef, bathypelagic, and a number of other special marine environments are largely excluded from discussion.

In the first two chapters we have attempted to describe the plankton community in terms of its composition and distribution of organisms. We felt that these descriptions were necessary for the reader to gain an introduction to the processes described in Chapters 3 to 5. Chapters 1 and 2 give some indication, therefore, of the complexity of both plankton distributions and the chemistry of plankton. Added to this complexity, however, are the results obtained by different scientists using different techniques in widely separated areas of the world. Consequently a synthesis of results reported in the literature is difficult. Instead we have tended to present data which offer the reader examples of biological oceanographic variability rather than trying to convince people of the acceptability of any one author's results in an absolute sense.

Chapter 3 deals with the primary formation of particulate material which is the beginning of the food chain in the pelagic environment. Feeding processes and the kinetics of food exchange in the pelagic food web are discussed in Chapter 4. In Chapter 5 an attempt is made to relate various processes into cycles which emphasize the interdependence of all processes in the sea.

In the final chapter we have given a number of examples of problems in the marine environment which we feel require the particular attention of the biological oceanographer. These have been chosen as representing areas in which there are also interdisciplinary interests between biological oceanographers and other professionals working in the marine habitat, including physical oceanographers and fisheries experts. The chapter is not intended to solve any problems but to suggest, by example, where there is a basis for the solution of such problems, using some of the information contained in the first five chapters.

Symbols used in equations and figures have presented us with a problem since aquatic biologists have tended to use the same symbol for several different entities. Thus R is commonly used to represent 'ration' and 'respiration'; P is used to represent the element phosphorus, photosynthesis and production. Where possible we have kept the common usage of the symbols and defined each one in the immediate context of its use. Where two processes are defined by the same letter in the same equation we have differentiated between the symbols used. In referring back to the literature we felt that the preservation of the popular symbols would be less confusing than introducing a large number of new symbols. Biological oceanographers have also tended to use different units for measuring the same parameters. Added to these differences is the fact that there is a lack of uniformity in the use of abbreviations for the same units. For example, light has been reported both in energy units and units of illumination, while the distance of one micron is sometimes abbreviated as 1 μ or 1μm. It is not our purpose to endorse any uniform use of units or abbreviations but it is our purpose to clarify the literature by showing how units can be converted and where any similarity in abbreviations may exist.

For more detailed coverage of specific subjects in biological oceanography, the reader will find a number of recent reviews in multi-author texts or in journals; also several books can be particularly recommended for coverage of the literature up to the date of their publication. The latter include "The Chemistry and Fertility of Sea Waters" by H. W. Harvey (Cambridge University Press, 1957); "Measuring the Production of Marine Phytoplankton" by J. D. H. Strickland (Queens Printer, Ottawa, 1960) and "Plankton and Productivity in the Oceans" by J. E. G. Raymont (Pergamon Press, Oxford, 1963).

Finally the authors would like to acknowledge the advice of Dr. G. L. Pickard, Director of the Institute of Oceanography, University of British Columbia and other members of the Institute who read the text prior to its publication. Many thanks are also extended to Dr. R. W. Eppley, Scripps Institution of Oceanography and Dr. G. H. Geen, Simon Fraser University, who were invited by the authors to criticise a draft edition of the text. In acknowledging the assistance of these scientists we do not hold them responsible for any errors or misconceptions which may have been included in the text. This work was in part supported by the National Research Council of Canada and by the Fisheries Research Board of Canada.

CHAPTER 1

DISTRIBUTIONS OF PLANKTON AND NUTRIENTS

1.1 TAXONOMIC, ENVIRONMENTAL
AND SIZE SPECIFIC GROUPS OF PLANKTON

Organisms which are unable to maintain their distribution against the movement of water masses are referred to as 'plankton'. Included in this group are bacterioplankton (bacteria), phytoplankton (plants) and zooplankton (animals). Generally all plankton are very small and in many cases, microscopic. However, relatively large animals, such as the jellyfish, are also included in the definition of plankton. Some plankters, including both plants and animals, are motile but their motility is weak in comparison with the prevailing movement of the water. Animals, such as fishes, which can maintain their position and move against local currents are known as 'nekton'. However, the division between plankton and nekton is not precise and some small fish, especially fish larvae, may be a part of the plankton community.

The biomass or weight of plankton or nekton per unit volume or area of water is referred to as the 'standing stock'; typical units used for standing-stock measurements are $\mu g/l$, mg/m^3, g/m^2, kg/hectare, etc., where the weight should be specified as referring to wet weight, dry weight, or carbon. The productivity of organisms is defined in terms of 'primary productivity', 'secondary productivity', and 'tertiary productivity'; units are the same as in standing-stock measurements when expressed per unit time (e.g. per hour, day, or year). Ideally, primary productivity represents the autotrophic fixation of carbon dioxide by photosynthesis; secondary productivity represents the production of herbivorous animals and tertiary productivity represents the production of carnivorous animals feeding off the herbivore population. However, these definitions are not precise since some plants may utilize growth factors, such as vitamins (auxotrophic growth), and others are capable of taking up organic substrates as a source of energy (heterotrophic growth). Thus the particulate material grazed by secondary producers may be derived from a variety of processes and include phytoplankton and bacterioplankton. Similarly many filter-feeding zooplankton which might be nominally classed as herbivores, may at times feed upon other small animals, such as Protozoa; thus the boundaries between components in the aquatic biosphere are difficult to define with the same precision as is used in chemistry or physics. Biological associations are better considered *in toto* as an ecosystem in which various components react with each other to a greater or lesser degree. Components of an ecosystem can be defined in terms of their taxonomy or chemistry; interactions between components can then be expressed quantitatively by empirical equations. Thus a phytoplankton standing

1

stock may be described as consisting of 10^6 cells per litre of a species, *Skeletonema costatum*, or as being represented by a chlorophyll *a* concentration of 1 mg/m^3, or in a trophic sense as a ration for zooplankton, such as is represented in eqn. (77), Section 4.2.1. Attempts to synthesize the components of biological production into simulated models of various ecosystems are currently being made, but the results of this work are at present largely experimental.

The bacterioplankton of the oceans have been discussed in some detail by a number of authors including ZoBell (1946), Kriss (1963) and Wood (1965). According to Wood (1965) the principal genera of bacteria represented in the oceans are the *Micrococcus*, *Sarcina*, *Vibrio*, *Bacillus*, *Bacterium*, *Pseudomonas*, *Corynebacterium*, *Spirillum*, *Mycoplana*, *Nocardia*, and *Streptomyces*. Among these genera are various morphological differences including coccoid, rod, and spiral forms. A large number of the most common bacteria are motile and gram negative. The bacterioplankton do not usually contribute significantly to the total biomass of particulate organic matter but in association with detritus (see Section 2.4) they may form an appreciable organic reserve during times of low phytoplankton density. Their role in the oceans is more important in recycling elements and organic material back into the food chain (see Fig. 37).

MacLeod (1965) has discussed the specific identity of marine bacteria as compared with bacteria from a terrestrial origin. His findings showed that marine bacteria have special requirements for inorganic ions which include a highly specific need for Na$^+$ and a partial need for halide ions which could be satisfied by either bromine or chlorine ions. Mg^{2+} and Ca^{2+} were also required, usually at concentrations higher than are normally needed for terrestrial bacteria.

Marine bacteria living at the sea surface are usually recognized as a specific community (e.g. Sieburth, 1971; Tsyban, 1971). In samples from the eastern Pacific ocean (between 10 and 30°N), Sieburth (1971) found the bacterial flora at the surface (bacterioneuston) were predominantly atypical pseudomonads having marked lipolytic and proteolytic activity. Tsyban (1971) describes the bacterioneuston from the Black Sea and northeast Pacific as having specific biochemical properties and consisting of brightly pigmented strains. The number of organisms at the sea surface increases during periods of wind and wave action and this has given rise to the question of whether bacterioneuston are accumulated largely through physical forces or through propagation (Marumo *et al.*, 1971).

Phytoplankton taxonomy has undergone frequent revision but for the purposes of our discussions, the "Check-list of British Marine Algae—2nd Revision" by Mary Parke and Peter Dixon (1968) has been followed in reporting on various taxonomic groups and species. The major classes of algae which contain representatives of the phytoplankton are as follows.

Taxonomic class	*Common name*
Cyanophyceae	(Blue–green algae)
Rhodophyceae	(Red algae)
Cryptophyceae	—
Dinophyceae	(Dinoflagellates)
Haptophyceae	—
Chrysophyceae	(Yellow–brown algae)
Xanthophyceae	(Yellow algae)
Prasinophyceae	—
Chlorophyceae	(Green algae)
Bacillariophyceae	(Diatoms)

Among these classes, the Bacillariophyceae, Dinophyceae, and Chrysophyceae are generally considered to be the most important in the sea; under some conditions the Cryptophyceae, Chlorophyceae, Cyanophyceae, and Xanthophyceae may be very abundant. The principal exception to the classification given above is that the Dinophyceae are sometimes included in the animal kingdom among the phylum, Protozoa.

The zooplankton include practically every major taxonomic group of animals, either for their entire life cycle or, for short periods, as in the case of the larval stages of some fish and molluscs. The major phylla represented in the following list have been taken in part from a much more detailed summary provided by Newell and Newell (1963).

Animal phylum	*Some representative among the zooplankton*
Protozoa	Oligotrich and Tintinnid ciliates, Radiolaria and Foraminifera
Coelenterata	Hydrozoa, Scyphozoa (Jellyfish)
Ctenophora	(Ctenophores)
Chaetognatha	(Arrow worms)
Annelida	(Polychaete worms)
Arthropoda (class) Crustacea	(Copepods, cladocerans, mysids, euphausiids, ostracods, cumaceans, amphipods, isopods)
(Sub-Phylum) Urochordata	(Salps and appendicularians)
Mollusca	(Heteropods)

In addition to the above, there are many large invertebrates having larval planktonic stages (e.g. polychaetes, crustaceans, gastropods, lamellibranchs, and echinoderms). Among vertebrates, fish eggs and larvae both occur as members of the plankton. Important commercial species which have a planktonic stage include the herring, anchovy, tuna, and bottom feeders, such as cod and plaice. From among the types of zooplankton listed above, by far the most important group are the Crustacea, and of these, the copepods are the most predominant.

Biogeographical distributions of plankton have been based on the very early recognition that specific environmental factors, such as light, temperature, salinity, and nutrient requirements, to some extent determined the occurrence and succession of species. Smayda (1958 and 1963) has reviewed a number of the terms used to describe plankton from similar environments. Plankton with a tolerance to a wide range of temperatures is described as 'eurythermal', while a narrow range of temperature tolerance is described as 'stenothermal'; similarly salinity (euryhaline and stenohaline), pH (euryionic and stenoionic), and light (euryphotic and stenophotic). A further classification of light response has been used to obtain a vertical separation of plankton communities into those inhabiting the 'euphotic' zone (where the net rate of photosynthesis is positive) as opposed to the 'aphotic' zone. Unfortunately, these words usually lack precise quantitative description but some knowledge of their meaning may be useful in reading other publications.

The terms 'oceanic' and 'neritic' have been used quite extensively in describing plankton associated with the oceans and with coastal waters, respectively. The classification may be particularly useful in reporting taxonomic data collected from commercial vessels, such as with a Hardy recorder as illustrated in Fig. 1. From this figure it is easy to see that certain "indicator species" belong to each region. Over large areas of ocean there may be several oceanic groups. Bary (1959 and 1963) defined such plankton distributions in terms of their temperature and salinity tolerances (called T–S–P diagrams). He emphasized that the importance of such diagrams was in showing the distribution of plankton in certain water bodies rather than the exact geographical location of the samples. Fager and McGowan (1963) used an elaborate mathematical technique to show probable affinities between zoo-

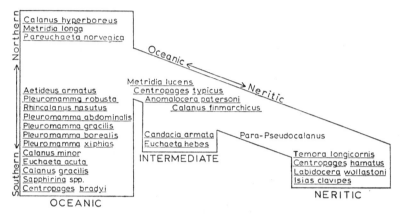

FIG. 1. Distribution series of copepods in the North Sea and the north-eastern Atlantic arranged in such a way that the distribution of each organism is most similar to those of the neighbouring organisms in the list (redrawn from Colebrook *et al.*, 1961).

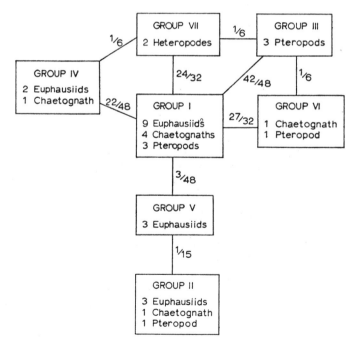

FIG. 2. Composition of zooplankton groups in the north Pacific. Fractions are the ratios of the number of observed species–pair connections between groups to the maximum number of possible connections; for example, there are six possible intergroup species pairs between group IV and VII but only one of these showed affinity at the 'significance' level used (redrawn from Fager and McGowan, 1963).

plankton species in the north Pacific. They classified the zooplankton into 6 groups and showed interrelationships between groups as illustrated in Fig. 2. Perhaps the most important aspect of all such groupings is that the occurrence of species in another group (or changes in the between group associations) may indicate changes in the ocean and coastal environments, such as might occur from a shift in the direction of a current. However, Fager and McGowan (1963) found that many of the usually measured properties of water (e.g. temperature, thermocline depth, etc.) were not closely correlated with differences in zooplankton abundance. From this it was concluded that the organisms reacted to a more complex interaction of known properties or to some environmental factors yet to be elucidated.

Phytoplankton species which produce resting spores or have a sedentary phase are known as 'meroplankton' as opposed to 'holoplankton' (Smayda, 1958). However, these terms were originally used to refer to zooplankton and in this sense 'meroplankton' refers to organisms which are only temporarily members of the plankton community (e.g. some bivalve larvae) while 'holoplankton' refers to a permanent member of the plankton community (e.g. most copepods). A general classification of plankton abundance based on availability of nutrients is used in describing waters as 'eutrophic', 'mesotrophic', and 'oligotrophic', in decreasing order of plankton abundance (see Hutchinson, 1969, for a further discussion of these terms). Plankton may be grouped by the depth zone in which they are found in the 'pelagic' or open-sea environment. These zones have received a number of different classifications but the simplest approximate definitions appear to be 'epipelagic' (0 to 150 m), 'mesopelagic' (150 to 1000 m), 'bathypelagic' (1000 to 4000 m), and 'abyssopelagic' (4000 to 6000 m) (see Hedgpeth, 1957, for further definitions). Plankton (or other particulate matter) produced within a designated ecosystem is referred to as 'autochthonous' while 'allochthonous' material is imported into the ecosystem. A large number of

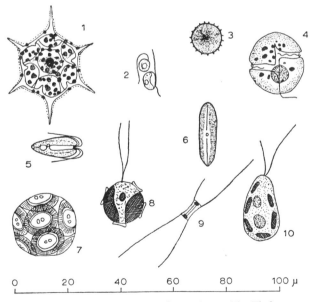

FIG. 3A. Examples of nanoplankton: flagellates [*Distephanus* (1), *Thalassomonas* (2), *Gymnodinium* (4), *Tetraselmis* (5), *Coccolithus* (7), *Pontosphaera* (8), *Cryptochrysis* (10)], diatoms [centrate (3), pennate (6), *Chaetoceros* (9)]. Redrawn from Wailes (1939), Cupp (1943), Fritsch (1956) and Newell and Newell (1963).

other groupings have been employed by systematists in describing plankton communities. In some cases these terms lack universal usage because of the specificity of their original definition; others have acquired common scientific usage while lacking a precise definition (e.g. see Smayda, 1958). Some attempt has been made to adopt universal definitions of certain terms, and the Report of the Committee on Terms and Equivalents (1958) has been followed where possible in this text.

From the point of view of food chain studies one of the most useful groupings for plankton and larger organisms is to consider all particulate material on a single size scale. The original definitions for the size grouping of planktonic organisms have been discussed by Dussart (1965), who suggested that plankton should be classified according to the scheme, 'ultra-nanoplankton' ($<2 \mu$); 'nanoplankton' (2–20μ); 'microplankton' (20–200μ); 'macroplank-

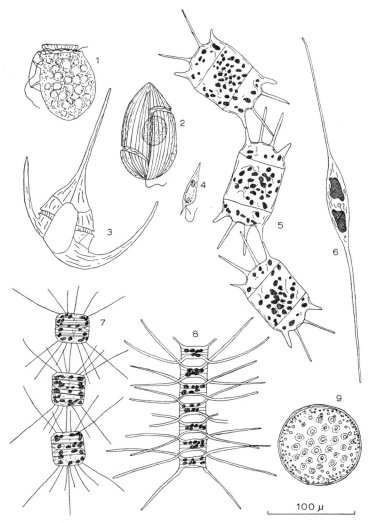

FIG. 3B. Examples of microphytoplankton: dinoflagellates [*Dinophysis* (1), *Gyrodinium* (2), *Ceratium* (3), *Prorocentrum* (4)], diatoms [*Biddulphia* (5), *Nitzschia* (6), *Thalassiosira* (7), *Chaetoceros* (8), *Coscinodiscus* (9)]. Redrawn from Wailes (1939) and Cupp (1943).

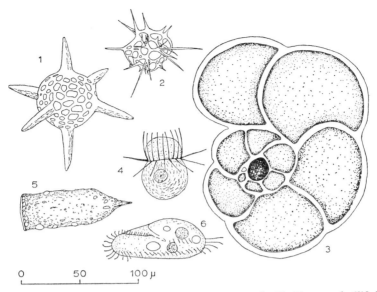

FIG. 3C. Examples of microzooplankton: radiolarians [*Hexastylus* (1), *Plectacantha* (2)], foramini-feran [*Pulvinulina* (3)], ciliates [*Mesodinium* (4), *Tintinnopsis* (5), *Amphisia* (6)]. Redrawn from Wailes (1937 and 1943) and Cushman (1931).

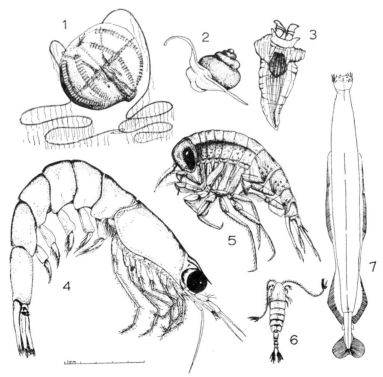

FIG. 3D. Examples of macro- and mega-zooplankton: ctenophore [*Pleurobrachia* (1)], mollusc pteropods [*Limacina* (2), *Clione* (3)], euphausiid [*Thysanoessa* (4)], amphipod [*Parathemisto* (5)], copepod [*Calanus* (6)], chaetognath [*Sagitta* (7)]. Redrawn from LeBrasseur and Fulton (1967).

ton' (200–2000 μ); 'megaplankton' (>2000μ)*. Illustrations of the types of planktonic organisms which might be considered in Dussart's size categories are shown in Fig. 3 A, B, C, and D. Thus the smallest phytoplankton represent most of the nanoplankton (Fig. 3A) while below this size group, the ultrananoplankton are chiefly represented by the bacterioplankton. The microplankton may consist both of large phytoplankton (Fig. 3B) and small zooplankton (Fig. 3C), while the macro- and megaplankton are generally represented by the largest zooplankton (Fig. 3D), including jellyfish and the larval stages of some fish. While these definitions were originally intended for grouping of organisms as measured under a microscope, it later became possible to obtain approximate size groupings of organisms using filters with different pore sizes. From the latter studies (e.g. Anderson, 1965; Teixeira *et al.*, 1967; Saijo and Takesue, 1965) it was generally shown that maximum photosynthetic activity occurred among the smaller phytoplankton species (*ca* 5–50 μ). The method of separation lacked precision, however, and has been criticized by Sheldon and Sutcliffe (1969) because the stated pore sizes of filters are not a good indicator of their effectiveness for separating size fractions of suspended particles. The use of the Coulter Counter® for analyzing size fractions of particulate material has been described by Sheldon and Parsons (1967a and b, and references cited therein). The operation of this instrument depends upon measuring the amount of electrolyte displaced as a particle moves through a small electric

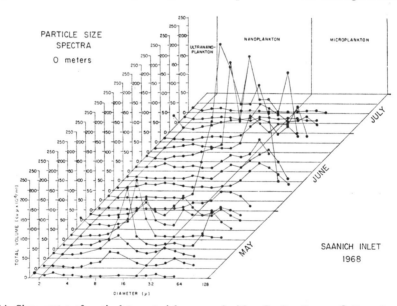

FIG. 4A. Size spectra of particulate material measured with a Coulter Counter® (from Parsons, 1969.

field. Thus the instrument measures particle volume; size in linear dimensions is then related to the diameter of a sphere equivalent in volume to the original particle. Size spectra of particulate material measured with a Coulter Counter® are shown in Fig. 4A. The figure illustrates differences in the size of particles (1 to 100μ) in the surface waters of a fiord on the coast of British Columbia. Fig. 4B shows an idealized representation of data that can be obtained by expanding the scale to include zooplankton. Unfortunately, at

* 1 μ, or micrometer (μm) ≡ 10^{-3} mm ≡ 10^{-6} m.

present there is no instrument of similar accuracy to a Coulter Counter® for measuring large particles. The size scale used in this spectrum is based on measuring the number of particles in size groups which progress by doubling the volume in each particle size category. The equivalent diameters obtained from these volumes give a grade scale (based on $2^{1/3}$) in which the size categories suggested by Dussart (1965) appear at equal intervals along the abscissa (e.g. Fig. 4B). The total biomass of material measured with the Coulter Counter®

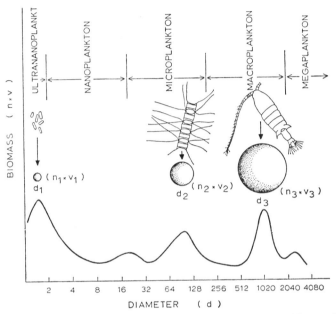

FIG. 4B. Particle spectrum representing biomass ($n \times v$) of material in different size categories determined by the diameter (d) of a sphere equivalent in volume (v) to the original particle, times the number of particles (n).

is statistically related to such parameters as the weight of particles, chlorophyll *a* concentration and particulate carbon (e.g. Zeitzschel, 1970). The advantages of using a continuous size spectrum for particle distributions in the sea are (i) the size group of plankters contributing most to the total standing stock of plankton can be readily identified as peaks in the spectrum, (ii) biomass diversity indices (Wilhm, 1968) can be calculated from the spectrum, and (iii) the growth increment, or grazing loss, of different size categories can be determined independently of the total biomass of phytoplankton, zooplankton, and other particles (Parsons, 1969; Parsons and LeBrasseur, 1970). However, particle size spectra *per se* do not relate to taxonomic groups and microscopic identification of the principal components in a plankton crop is recommended when using this technique. Also the results include all detrital particulate material, and special methods are sometimes necessary to differentiate between detritus and growing cellular material (Cushing and Nicholson, 1966).

Attempts to measure zooplankton biomass in different size categories have been suggested by Chislenko (1968) who has produced nomographs giving the different weights from the length measurements of a large number of different zooplankton shapes. More recently an automated apparatus for measuring the length of zooplankton has been described by Fulton (1972).

1.2 DIVERSITY

The diversity of a plankton community may be expressed using data on the number of species present, the distribution of biomass, the pigment composition, or a number of other parameters which are easily measured properties of plankton. From a heuristic approach, an index of diversity may be used in the same way as other environmental parameters, such as temperature and salinity, to characterize the environment. There are both theoretical and empirical bases for the use of specific diversity indices, but as Lloyd *et al.* (1968) have stated, "which one is 'best' depends upon which one proves in practice to give the most reliable, surprising ecological predictions and the greatest insight".

The simplest expression of diversity is to determine the percentage composition of species in a sample; the more species making up the total, the greater is the diversity of the organisms. However this value is almost wholly dependent on the total number of individuals (N) and is therefore unsatisfactory as a diversity index. From some of the earliest quantitative studies in ecology it was recognized, however, that a relationship existed between the number of species in a population (S) and the logarithm of the total number of individuals (N), so that the simplest diversity index (d) can be expressed as:

$$d = \frac{S}{\log_{10} N}.$$ (1)

This value will be very small under conditions of a plankton bloom and generally high in tropical plankton communities. A better expression which reduces to 0 when all the individuals are from the same population was given by Margalef (1951):

$$d = \frac{S-1}{\ln N}.$$ (2)

Margalef (1957) introduced the idea that the 'information content' could be used as a measure of diversity in a plankton sample. Thus the diversity of a collection containing a total of N individuals and $n_1, n_2, \ldots n_i$ individuals of each species can be written as:

$$\frac{N!}{n_1!\, n_2!\, \ldots\, n_i!},$$ (3)

and information (H in 'bits')* per individual as:

$$H = \frac{1}{N}\, \log_2 \frac{N!}{n_1!\, n_2!\, \ldots\, n_i!}.$$ (4)

The information content as expressed above, eqn. (4), can be interpreted as the degree of uncertainty involved in predicting the species identity of a randomly selected individual. If N is large and none of the n_i fraction are too small, information content per individual (H' in bits) can be approximated from the expression:

$$H' = -\Sigma p_i \log_2 p_i,$$ (5)

* Information is commonly expressed in 'bits' when using $\log_2$, or 'nats' when using $\log_e$. The following conversion may be useful: $\log_{10} = \log_e \times 2 \cdot 303$ and $\log_2 = \log_e \times 1 \cdot 443$.

where $p_i = n_i/N$ and is the proportion of the collection belonging to the i th species (Shannon and Weaver, 1963). Under some circumstances eqn. (5) will be more easily determined than eqn. (4) but Lloyd et al. (1968) have provided examples and tables for the solution of both equations for values of N from 1 to 1050. Margalef (1961) made a statistical comparison between the diversity of plankton samples calculated from the diversity index [eqn. (2)] and the theoretical diversity [eqn. (4)]. There was a highly significant correlation between the two although there was a difference in the regression line depending on whether diversity was determined for a diatom or dinoflagellate population. Using eqn. (5) Lloyd and Ghelardi (1964) related diversity to the maximum possible value for a given number of species if they were all equally abundant. This term was called the 'equitability' (ε) and was expressed as the ratio of H' to a theoretical maximum (M) for the same number of species, where $n_1 = n_2 = \ldots n_i$. A table of M values for 1 to 1000 species is given by the authors. The value of ε in describing a collection may be more useful if units of biomass rather than number are employed to determine diversity (e.g. Wilhm, 1968).

"Equitability", as defined by Lloyd and Ghelardi (1964) is the opposite of "dominance" which expresses the most abundant species in a population. Hulbert et al. (1960) expressed the dominance of a plankton community as the ratio of the concentration of the most abundant species to the total cell concentration. Also if the presence of one species in a population is nearly always accompanied by another species, the amount of information gained is small. This can be expressed as the "redundancy" (R) in terms of the equation (Patten, 1962a):

$$R = \frac{H_{\max} - H}{H_{\max} - H_{\min}}, \tag{6}$$

where $H_{\max}$ is the diversity when the species are equally distributed and $H_{\min}$ is the diversity when all the individuals belong to one species. The value R, varies between 0 and 1, and is also partially an index of "dominance".

The general use of diversity of the type shown in eqns. (4) and (5) above has been discussed by Pielou (1966) who points out that the diversity of a sample should not be regarded as the diversity of a larger population from which it was obtained; the sample itself may, however, be treated as a population and defined. Patten (1959) discusses the absolute diversity of an aquatic community in terms of its information content. As an approximation he calculated that in a Florida lake, the community could be described in terms of 3×10^{24} bits/cm^2/year. If the average information content of a printed page is 10^4 bits, it is apparent that the amount of information required annually to describe the Florida lake community is many orders of magnitude larger than the information contained in the largest libraries! Thus diversity, as discussed in this section, is a property of the entity from which the data are collected and not of the whole environment.

The species diversity indices discussed above [e.g. eqn. (2)] are dependent to some extent on the size of the sample, especially for small numbers ($N < 100$). In a method used by Sanders (1968), however, samples of benthic fauna from the same environment, ranging in size from 35 to 2514 individuals, showed no tendency for smaller samples to be less diverse. The technique is described as the 'rarefaction' method and it depends on determining the shape of the species abundance curve rather than obtaining an expression for the absolute number of species per sample. The method is demonstrated as follows using a hypothetical plankton sample shown in Table 1. The problem is to generate a species abundance curve for different sample sizes (Fig. 5). If a sample size of 20 individuals is selected, this will represent 100% of the individuals and each individual will represent 5% of the sample.

In the original sample, 4 species represented 5% or more of the total or 89% of the sample
(Table 1). Therefore it is assumed that each of these species will be in the reduced sample
which leaves 11% of the sample for the remaining 6 species; since none of them form more
than 5% of the total, none is likely to be represented by more than one individual. Thus
the total number of species present in a sample of 20 is $4 + 11/5 = 6 \cdot 2$. Similarly the number
of species in a sample of 10 individuals is 5 and so on; the shape of the species abundance
curve generated by Table 1 is shown in Fig. 5. Also, two curves are shown with lower
diversity (bottom) and higher diversity (top). The author emphasizes that in order to use
this method, samples must be taken by the same sampling technique, from similar environments (i.e. it would be inappropriate to compare the diversity of attached algae and phytoplankton) and curves may not be extrapolated. The curves (Fig. 5) can not have confidence
limits applied to them but curves from samples with similar diversities emerge as a family
of curves which strongly indicate that the diversity differences are real.

TABLE 1. THE RAREFACTION METHOD OF CALCULATING DIVERSITY—A
HYPOTHETICAL PLANKTON SAMPLE CONSISTING OF 100 INDIVIDUALS AND
10 SPECIES (A LARGER EXAMPLE IS GIVEN BY SANDERS, 1968).

Rank of species by abundance	Number of individuals	% of sample	Cumulative sample %
1	45	45	45
2	31	31	76
3	8	8	84
4	5	5	89
5	4	4	93
6	3	3	96
7	1 each	1 each	100
TOTAL	100	100	

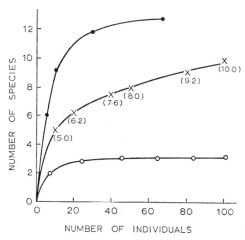

FIG. 5. Plot of number of species at different population levels using the rarefaction diversity
method for different plankton samples. (×—× curve generated from Table 1; ○—○ low
diversity sample; ●—● high diversity sample).

Changes in the diversity index of samples from a plankton community are shown in Fig. 6. From these data Margalef (1958) recognized 3 stages of succession. Stage 1 was typical of turbulent waters in which a few species survived and in which there was an occasional bloom of diatoms; stage 3 was characteristic of highly stratified waters in which there was a mature phytoplankton crop and a high diversity following nutrient depletion. Stage 2 was characteristic of inflowing waters which may have transported allochthonous species into the area of study, thus increasing the diversity of organisms present in any sample. Hulbert *et al.* (1960) studied changes in species diversity in the Sargasso Sea and recognized a succession of three species groups. These consisted of a sparse population with a normal

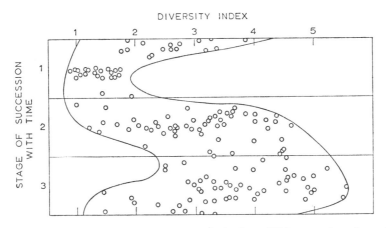

FIG. 6. Stage of succession dependence of diversity index [eqn. (2)] in a number of samples from the surface waters at the bay of Vigo (redrawn from Margalef, 1958).

distribution of abundant and rare species, a winter period in which a single species was dominant over all other species, and a period of thermal stratification in which dominance was shared by several species. The most extensive field tests of various diversity indices have been carried out by Travers (1971) in the Mediterranean. From this study the author concluded that the degree of maturation and of organization of an ecosystem can be appreciated by means of several diversity indices. From the use of different indices he concluded that a diversity index based on plankton pigments (Margalef, 1965) was a poor method, especially where plankton levels were low; diversity indices based on information theory were considered the best measure of structure although less laborious calculations of diversity can often be used [e.g. eqn. (2)].

1.3 SPATIAL DISTRIBUTIONS

1.3.1 STATISTICAL CONSIDERATIONS

In carrying out a series of replicate analyses on a single, well mixed sample of sea water, small differences in the values obtained may be attributed to analytical technique. Such errors are caused by a lack of instrument reliability, sub-sampling, and slight variations in the way an individual analyst repeats each analysis. It may be assumed that these errors are randomly distributed and that if a large number of replicate analyses are made on one sample, the mean and standard deviation (s) of the analysis can be determined. The precision of the method can then be expressed with a 95% confidence limit for n determinations, as;

$$m \pm 2s/\sqrt{n}, \tag{7}$$

where m is the mean of the replicate samples. The principal exception to the use of these statistics for analytical techniques is in the counting of plankton from a settled volume of sea water. In this case the distribution of plankton cells may not be random and special methods may have to be used in order to determine the degree of contagion (Holmes and Widrig, 1956).

The collection of samples of seawater, or plankton from the ocean, introduces much larger differences between replicate samples than can be ascribed to analytical errors alone. Thus Cushing (1962) has summarized a number of reports and showed that in calm weather, the % variability (expressed as the coefficient of variation $s/m \times 100$) of the number of 3 species of plankton in individual hauls varied from 15 to 70%; under conditions of rough weather this variability was increased up to 300%. Cassie (1963) estimated that the coefficient of variation for large samples is most often in the range of 22 to 44%, with obvious exceptions being made for rough weather, or highly stratified environments. Wiebe and Holland (1968) have summarized data on the 95% confidence limits for *single* observations of zooplankton abundance; the range for most data was between *ca* 40 and 250%.

While the use of sampling gear itself may contribute a small amount of variability to ocean sampling (Cushing, 1962, assigns *ca* 5% variability to gear operation), the principal cause of variability in replicate samples is due to the non-random or patchy distribution of plankton, and other non-conservative properties, such as the concentration of nutrients. The mechanisms leading to these differences in spatial abundance are many and diverse. They include the physical accumulation of particles by the vertical and horizontal movement of water masses (e.g. divergence and convergence), differences in growth rates of individual plankters, and nutrient uptake and predation patterns of the food chain. These processes are discussed at other points in the text (see Section 1.3.5) and the following discussion (primarily from Cassie, 1962a) deals only with the extent and not the cause of distributions.

In random distributions, two or more samples of seawater of a given volume are equally likely to contain the same organism. The expected distribution of samples with n_1, n_2, n_3 etc. individuals is given by successive terms of the binomial expansion

$$(q+p)^k,$$

where k is the maximum number of individuals a sample could contain, p is the probability of an organism's occurrence and $q = 1-p$. The population mean (μ) and variance (σ^2) of a binomial distribution are

$$\mu = kp \quad \text{and} \quad \sigma^2 = k\,pq$$

from which

$$\sigma^2 = \mu - \mu^2/k. \tag{8}$$

Since for plankton in a seawater sample $k \rightarrow \infty$, the variance become equal to the mean.

$$\sigma^2 = \mu. \tag{9}$$

The above relationship expresses a special case of the binomial distribution, in which the probability of an organism occurring ($p = \mu/k$) is small; this is known as the Poisson distribution and the experimental value of the variance (s^2) and mean (m) for replicate plankton collections can be expressed as s^2/m and used to determine if the plankton are randomly or "over-dispersed". Theoretically if the value σ^2/μ is greater than 1, the population will be over-dispersed. However, in practice if s^2/m is calculated for a series of samples their distribution can be represented as $\chi^2/N-1$; thus with 20 samples (19 degrees of freedom), s^2/m should be less than $30\cdot14/19 = 1\cdot6$ for a 95% probability that the organisms are distributed randomly (Holmes and Widrig, 1956). However in most cases involving the collection of plankton over an area, the value s^2/m will be significantly greater than 1; thus the ratio can be used as a dispersion coefficient (Ricker, 1937) which along with other parameters (e.g. diversity indices) may be useful in characterizing a body of water.

When populations are over-dispersed the presence of one organism in a sample increases the probability of additional organisms of the same species occurring in the same sample. This is the opposite of a binomial distribution and it can be expressed theoretically as the negative binomial distribution which is given by an expansion of the expression

$$(q-p)^{-k}$$

where $q = 1+p$. The variance (σ^2) is

$$\sigma^2 = \mu + \mu^2/k. \tag{10}$$

As Cassie (1962b) has pointed out, since p and k are negative, they cannot have the same meaning as they did in the binomial distribution. In particular, however, k appears as a useful parameter for expressing the degree of patchiness, or contagion, in a population. Cassie (1962a) has given an estimate of $1/k$ as $\hat{c}$, where

$$\hat{c} = \frac{s^2 - m}{m^2}, \tag{11}$$

s^2 and m being the sample variance and mean, respectively. The expression [eqn. (11)] was used by Cassie (1959) as a coefficient of dispersion and he concluded that $\hat{c}$ was better than s^2/m, since $\hat{c}$ was not strongly correlated with the mean. This allows for a comparison of dispersion to be made between samples with different means.

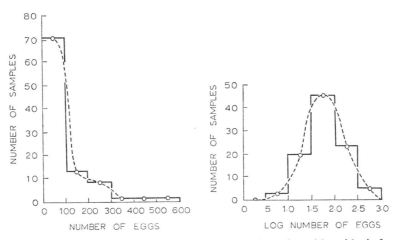

FIG. 7. Frequency distribution of pilchard egg counts in arithmetic and logarithmic forms (redrawn from Barnes, 1952).

In practice it may be found easier to establish an empirical relationship than to fit raw data to a theoretical distribution. Barnes (1952) and Cassie (1962a) have discussed transformations which may be suitable for marine biological data. The most convenient transformation is to convert the raw data to logarithms and the transformed frequencies may then have a log-normal distribution; an example of transformed plankton data is shown in Fig. 7. The mean (m') of the transformed data is the geometric mean and after taking antilogarithms, the value m' will be less than the arithmetic mean, m. Variability about the mean can be expressed as the 'logarithmic coefficient of variation' (Winsor and Clarke, 1940);

$$V' \text{ per cent} = 100\,(10^{s'} - 1), \tag{12}$$

where s' is the standard deviation calculated from the logarithms of the raw data and V' is the logarithmic coefficient of variation. The advantage of using V' has been illustrated by Cassie (1968) who considered two samples with means of 100 and coefficients of variation V (normal distribution) and V' (log-normal distribution) both of 110%. The mean with one standard deviation for the normal distribution was,

$$100 \pm (1 \cdot 10 \times 100) = -10 \text{ to } 210$$

and for log-normal

$$100 \overset{\times}{\div} 2 \cdot 1 = 48 \text{ to } 210.$$

The negative lower limit in the arithmetic range above is meaningless in the context of a normal distribution.

In a detailed study of spatial heterogeneity, Platt *et al.* (1970) separated the sources of variation in a single analysis as follows:

$$\sigma_T^2 = \sigma_0^2 + \sigma_1^2 + \sigma_2^2, \tag{13}$$

where σ_T^2 was the total variance, σ_2^2 was the variance due to real differences between stations, σ_1^2 was the variance between repeated samples taken at the same station and σ_0^2 included sub-sampling and other analytical errors. The authors found that σ_0^2 and σ_1^2 were about the same size and accounted for *ca* 10% of the variance each. Real differences between stations were generally much larger, both in time and space. Thus the log-coefficient of variation for a single chlorophyll *a* observation in a near-shore community increased rapidly from 14% at 0·625 sq. miles, to 70% at 1 sq. mile, thereafter remained relatively constant out to 4 sq. miles. During a study of temporal variations over a period of 5 weeks, the log-coefficient of variation varied from 11 to 111% (mean 42%) at 9 stations covering 13 sq. miles in the same location. During this period phosphate varied from 10 to 60%. Rapid temporal variations in V' sometimes occurred over a few days and ranged from 21 to 45% in the case of chlorophyll, and from 32 to 64% for phosphate. Since this study was conducted under relatively ideal weather conditions, the values quoted may be considered to be representative of maximum σ_2^2 values, but minimal σ_0^2 and σ_1^2 values. Thus Cassie (1962b) showed that patchiness was greatest in calm seas and least at times of turbulent mixing, such as during storms; conversely the operation of gear over the side at a single point, as well as laboratory analyses on board vessels, becomes much more difficult during rough weather and consequently the terms σ_0^2 and σ_1^2 may be expected to increase under such conditions.

Most of the discussion above pertains to overall descriptions of the patchiness of plankton communities in terms of plankton numbers, chlorophyll *a* or nutrients. Quite a different distributional problem arises when a description is required of the total number of species that might define the fauna of a particular area. The question is then a matter of how large a sample must be filtered in order to include all the species. In experiments conducted while following a current drogue, McGowan (1971) showed that for three groups of plankters (fish larvae, molluscs, and euphausiids) the amount of water to be filtered varied with the group. In the case of euphausiids there was no increase in the number of species with the volume of water filtered over the range 128 to 1510 m³, but for fish larvae and molluscs the amount of water to be filtered in order to include all species was in excess of 10,000 m³. However the relationship between the number of species and the volume filtered was logarithmic; for example *ca* 80% of the fish larvae species were found in *ca* 1000 m³ of water. The amount of water filtered by a plankton net can be increased either by increasing the size of the net or the length of the tow. Wiebe (1971) studied the precision of replicate tows with different nets and towing distances and his conclusion was that the length of the towing distance was considerably more important in determining precision than the size of the net.

1.3.2 AREAL DISTRIBUTIONS

A large amount of data has been collected on the areal distribution of plankton and nutrients. These data are generally the result of observations carried out at points along the cruise track of a research vessel; consequently while the data are useful for surveys of very large areas (e.g. seas, oceans, and the hydrosphere) they are of very little use in trophic studies. The latter subject is discussed later in the text, but for the present it must be apparent that plankton distributions mapped from samples collected miles apart are probably not representative of the food supply for a larval or juvenile fish, which may travel less than 100 m in a day. Thus small-scale plankton distributions are particularly important in assessing the food supply and hence, in part, the survival of very young fish.

It has been recognized from the time of the earliest explorers of the hydrosphere that plankton may sometimes occur in dense swarms or blooms (see Bainbridge, 1957, for historical references). Scientific observations (e.g. Barnes, 1949; Barnes and Marshall, 1951) showed that in general planktonic organisms were more often clumped or aggregated than randomly distributed. For example, Cassie (1959) showed in a study on the occurrence of plankton over a distance of 1 m that the distribution of the diatom *(Coscinodiscus gigas)* was non-random. The problem of collecting detailed samples over appreciable distances was solved in 1936 for larger plankters with the invention of apparatus which could be towed behind ships and continuously collect plankton on a slowly moving fine mesh belt (Hardy, 1936). An adaptation of this apparatus (Longhurst *et al.*, 1966) for studying micro-distributions of plankton has been particularly important for trophodynamic studies. Using this apparatus Wiebe (1970) was able to show areal patchiness in the distribution of zoo-plankton species over distances of less than 20 m. Some of the results obtained by Wiebe are shown in Fig. 8. From these data it is apparent that there is both a small-scale and a larger-scale patchiness in the distribution of the species reported. Due to the mechanical ability of the apparatus, the minimum distance over which the plankton patches could be detected with Longhurst–Hardy recorder was *ca* 14 m.

The most extensive descriptions of large-scale plankton species distributions are contained in reports from the Oceanographic Laboratory, Edinburgh (published in Bulletins of Marine Ecology). Zooplankton data are obtained from samples collected with Hardy plankton recorders towed behind commercial vessels (Hardy, 1936); an example of the descriptive data is given in Fig. 9. The numbers of zooplankton are reported as averages by rectangular

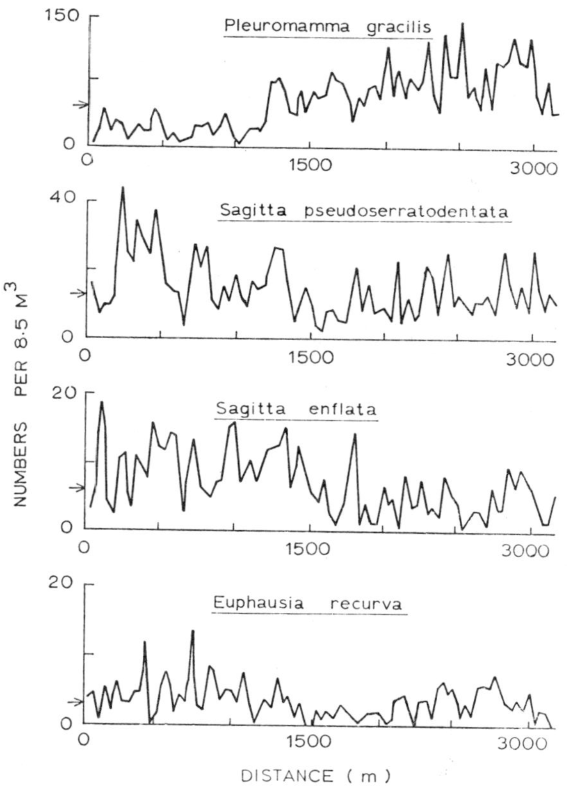

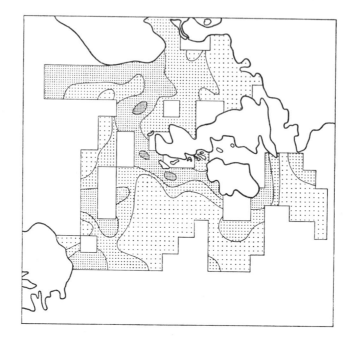

Number per sample   < 30   30 - 100   100 - 300   300 <

Fig. 9. The distribution of *Calanus finmarchicus*, stages V and VI from data obtained with a Hardy plankton recorder. Data show the number of animals per sample; blank rectangles indicate insufficient data (redrawn from Colebrook *et al.*, 1961).

Fig. 8. Plots of abundance versus distance illustrating the presence of large scale patchiness on which is superimposed smaller scale patchiness (redrawn from Wiebe, 1970).

sub-divisions for the North Sea and the Atlantic approaches to the British Isles. Similar data are collected for the larger phytoplankton species which are reported as a percentage incidence for each species.

Some data on regional differences in the plankton on an oceanic scale have been reported on for all of the world's oceans. For example, Omori (1965) has defined three oceanic regions in the north Pacific based on the distribution of three species-groups of copepods. These are 1) a cold offshore water region characterized by *Calanus plumchrus–C. cristatus*, 2) a warm offshore water region associated with *Calanus pacificus* and, 3) a neritic water mass region represented by *Pseudocalanus minutus–Acartia longiremis*. The latter region is oceanographically very complex and large differences in plankton concentrations are encountered on oceanic approaches to neritic environments. Additional information on the oceanic distribution of certain plankton species can be obtained from a study of sediments. For example, in the case of coccolithophores, the calcium carbonate coccoliths are often preserved both in the surface sediments, and in fossil remains. McIntyre and Bé (1967) have used this technique to describe species-specific zones of coccolithophore production in the Atlantic Ocean and similar maps have been drawn to show the distribution of diatoms and planktonic foraminifera.

Various attempts have been made to summarize productivity data on a global scale. Koblentz–Mishke *et al*. (1970) have reported primary productivity data for the world's oceans, based on a review of a large number of reports and a modified version of their original figure has been redrawn in Fig. 10. From these results it is apparent that over large

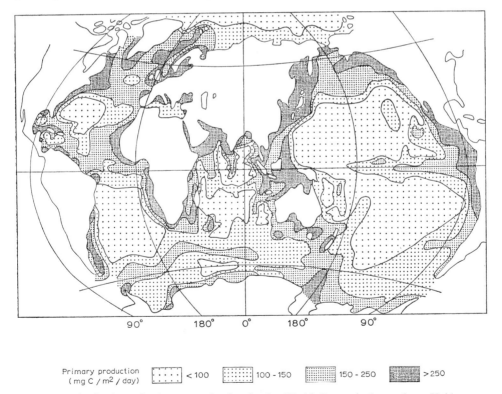

FIG. 10. Distribution of primary production in the World Ocean (redrawn from Koblentz–Mishke et al., 1970).

areas of the Pacific and Atlantic oceans, primary production is relatively low but that higher primary productivities are generally found in the proximity of land masses. There are some exceptions to this, such as where the South Equatorial current in the Pacific Ocean causes a band of relatively high primary productivity to occur along the equator.

Variations in the amount of plankton production in the pelagic environment may influence levels of benthic production. The chief factor in determining the level of benthic production in any area appears to be the depth of the water column. In reviewing this subject Rowe (1971) has presented data showing a statistical relationship between the $\log_{10}$ biomass of benthic animals and the depth of the water column in different environments. However, the second most important factor has been found to be areal differences in the level of primary productivity in the overlying water. Thus the sinking of phytoplankton from the surface layers generally contributes to the productivity of benthic animals; an exception to this relationship is apparent, however, under conditions where the bottom water has become anaerobic and the addition of further organic matter helps to maintain a lack of benthic fauna rather than assuring its existence.

The development of automated analysers has greatly assisted in the description of nutrient distributions in the ocean. An illustration of nitrate distribution obtained with an Auto-

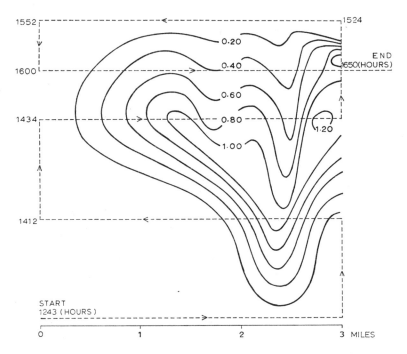

FIG. 11. Nitrate (μg at/l) at the surface off Punta Colnett Baja California (29°57'N, 116°20'W) 9 July, 1965. — nitrate concentration; - - - cruise track; redrawn from Armstrong et al., 1967.

analyser® during 8 transects of an approximate 10 sq. mile area is shown in Fig. 11. The illustration has been chosen to show changes in nutrient concentration in an area where nutrient depletion was general but in which there was some upwelling. Over larger areas of ocean, trends in nutrient concentration can be seen which are generally larger than the

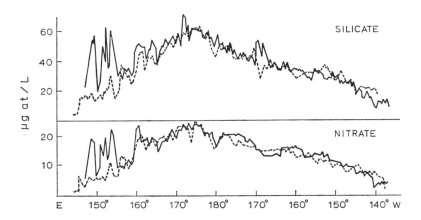

FIG. 12. Silicate and nitrate at 3 m as measured along an eastward and westward cruise track in the subarctic Pacific Ocean during April, 1969 (—— westbound, --- eastbound, Victoria, B.C. to Tokyo. Redrawn from Stephens, 1970).

small scale differences shown in Fig. 11. This is illustrated in Fig. 12 for nutrient observations carried out with an Autoanalyser® on two cruise tracks across the Pacific Ocean. The greater variability in the results in the western Pacific compared with the Gulf of Alaska, reflects the mixture of the two very different water masses (the Oyashio and the Kuroshio Currents) off the coast of Asia.

1.3.3 VERTICAL DISTRIBUTIONS

Until quite recently, most vertical profiles of biological parameters were made either with water bottles, which collected samples from discrete depths, or with plankton nets designed to open over some depth interval. The use of automated sampling gear, as well as recent advances in echo-sounding equipment, have greatly improved the data which are now being collected on vertical distributions. Strickland (1968) made direct comparisons between nutrients and chlorophyll *a* as measured in samples pumped from 0 to 75 m and the same data represented by a standard hydrographic cast. Differences between profiles integrated from standard casts and continuously recorded data were particularly marked in the case of chlorophyll *a*; for example, the chlorophyll *a* peak at *ca* 20 m in Fig. 13 measured 2·9 mg/m³ when detected in pumped samples using a fluorometer but was only 1·3 mg/m³ according to an integrated curve based on bottle casts at standard depths. Thus the total amount of chlorophyll *a* per m² integrated from a bottle cast or a continuous profile also tends to be different; to some extent, however, these errors are smoothed out and variations in chlorophyll *a* per m² were found by Strickland (1968) to be less than 25%. Nutrient analyses carried out at the same time showed less variability than the chlorophyll *a* data; nutrient data for integrated bottle and pump samples (i.e. per m²) generally differed by less than 10%.

The vertical distribution of chlorophyll in the sea generally shows a maximum which may sometimes be found near or at the surface and at other times, at or below the apparent euphotic depth (Steele and Yentsch, 1960). A deep chlorophyll maximum appears to be a

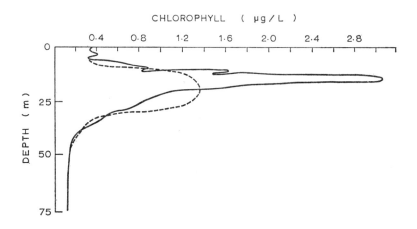

FIG. 13. Chlorophyll profiles integrated from standard bottle casts (– – –) and continuously recorded (—) data (redrawn from Strickland, 1968).

seasonal feature of summer vertical profiles as far north as 45 to 50° in both the Atlantic and Pacific Oceans. Anderson (1969) found the chlorophyll maximum off the Oregon coast at *ca* 60 m was formed by photosynthetically active cells which were apparently adapted to very low light intensity. Hobson and Lorenzen (1972) showed that chlorophyll maxima were associated with pycnoclines which occured at various depths in the Atlantic Ocean and Gulf of Mexico. Further it was apparent that increased concentrations of microzoo-plankton were associated with the chlorophyll maxima.

Large differences in the concentration of zooplankton at specific depths have been encountered using the Longhurst–Hardy recorder (Longhurst *et al.*, 1966), which collects samples continually by integrating catches over very short intervals of *ca* 10 m. An example of the zooplankton concentrations down to 400 m as measured with this apparatus is shown in Fig. 14A. The approximate 6-fold increase in zooplankton biomass at 300 m is a unique feature of this profile which would have been difficult to observe using data collected with a conventional plankton net. A similar result can be obtained with high frequency echo sounders as is illustrated in Fig. 14B from Barraclough *et al.* (1969). In this example the presence of echo-sounding material is indicated as a discrete band on the echogram. When the layer was sampled with conventional nets, however, the discrete accumulations of zoo-plankton appear more as a smoothed maximum in concentration. This is due to the gener-ally unavoidable collection of zooplankton at intermediate depths while nets are being lowered and brought up from specific sampling depths. One solution to this problem is the use of specially designed nets (e.g. the Clarke–Bumpus sampler) which can be made to open and close at the beginning and end of a specific sampling period. However, these samplers still represent integrated concentrations for the distance over which the net is towed when open.

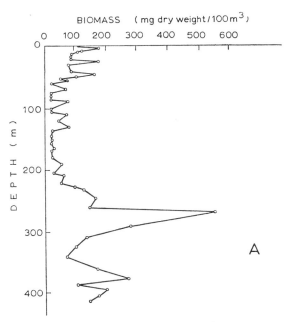

FIG. 14A. Vertical distribution of zooplankton biomass in the eastern Pacific Ocean; data obtained using a Longhurst–Hardy recorder (redrawn from Longhurst et al., 1966).

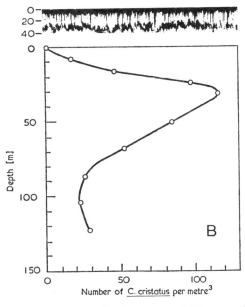

FIG. 14B. Depth profile of *Calanus cristatus* showing the actual echogram obtained with a 200 KHz recorder (top) and depth samples (○) using Miller nets (bottom). (Redrawn from Barraclough et al., 1969.)

A number of detailed studies have been carried out on sound scattering layers (for review, see Hersey and Backus, 1962); in particular there are many reports on deep scattering layers (DSL) which are usually found between 100 and 500 m in the oceans. Animals which have been found in scattering layers include squids, euphausiids, fish, and certain siphonophores. Although the types of animals which occur at discrete depths in the ocean may vary, it is probable that sound scattering with low frequency sounders (12-kHz) is only caused by certain specific animals, including large fish and particularly animals containing gas bubbles. Data from echograms, Longhurst–Hardy recorders, and closing nets all indicate that species of zooplankton and nekton generally occur over quite limited depth ranges and the general classifications of organisms by depth zone (e.g. epipelagic, mesopelagic etc. Section 1.1) may be employed in referring to the vertical distribution of animals. In some cases, however, animals may migrate vertically over distances of up to 1000 m in one day; in these cases specific depth location has little meaning.

The distributions of other biologically important materials generally show less patchiness with depth than occurs with the animals discussed above. Holm–Hansen *et al.* (1966) made detailed studies of a depth profile of nutrients (including nitrates, phosphates, ammonia, dissolved organic nitrogen, and Vitamin B_{12}) and particulate material (including particulate carbon, nitrogen, and phosphorus). Analyses were performed down to 1300 m off the coast of southern California. High concentrations of particulate materials were generally confined to the top 50 m euphotic zone; below 50 m the concentration of particulate carbon, nitrogen and phosphorus was about one fifth of the surface values and there was some indication of a further gradual decrease in particulate material to a tenth of surface values below 1000 m. One exception to these results was a high concentration of particulate carbon and phosphorus at 600 m in the profile. Nutrients (nitrates, phosphates, and Vit. B_{12}) showed the opposite distribution, being high in water below 50 m and low at the surface. These results are rather similar to a great many less-detailed descriptions of vertical profiles reported in the literature. From such descriptive data alone it is difficult to determine whether differences on a vertical axis are due to properties of the water column, such as vertical migrations, or if vertical differences are caused by lateral (Section 1.3.2) or temporal (Section 1.3.4) changes. Vertical differences in a water mass over a considerable lateral distance are illustrated in Fig. 15 for the distribution of particulate carbon along a transect in the Gulf of Aden. In this profile there is an apparent high concentration of particulate material between 400 and 600 m and another increase in the deep trench at 2000 to 3000 m. Similar differences in depth distribution of organic material were found by Menzel (1964) in the western Indian Ocean. On the other hand, Menzel (1967), reporting on the depth distribution of both dissolved and particulate organic carbon, showed that there was a relative constancy in the particulate (15 ± 5 μg/l) and dissolved (0.6 ± 0.1 mg/l) carbon in the tropical Atlantic, with slightly lower but also constant values for the tropical Pacific ocean, below 200 m. Some differences in technique, both in the collection of samples and analyses (see Section 1.3.1) may account for these differences in observations; alternatively they represent real differences in vertical distributions in different oceans which will require further examination and explanation.

A number of investigators have attempted to determine how much of the particulate material below euphotic zone is living. Thus apart from the obvious accumulations of certain animals at depths all the way to the bottom, the question arises whether there are appreciable quantities of unicellular organisms on which deep water filter feeders may exist without migrating. From studies on the distribution of ATP in the ocean, Holm–Hansen (1969a) concluded that the living component of deep water microscopic particulate

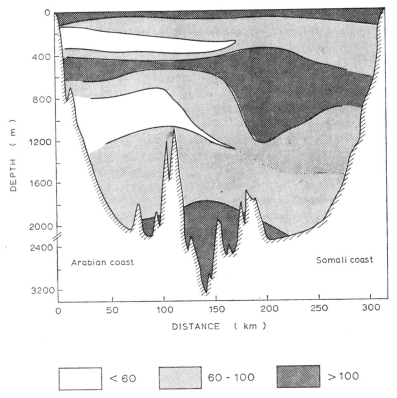

FIG. 15. Distribution of particulate carbon (μg/l) in the Gulf of Aden (redrawn from Szekielda, 1967).

material was only about 0·5 μg C/1, or 3% of Menzel's (1967) average value for deep water particulate carbon. From microscopic examinations, it has become apparent that pigmented cells, as well as bacteria, exist at great depths in the ocean although the role of these cells in the bathypelagic and abyssopelagic food chains has yet to be determined (Bernard, 1963; Fournier, 1966 and Hamilton et al., 1968).

The vertical distribution of the bacterioplankton has been studied by a number of authors (e.g. ZoBell, 1946; Kriss et al., 1960 and Sorokin, 1964b). Methods used for determining the presence of bacteria have led to very different assessments of the total biomass of bacterioplankton in the water column. Thus Kriss et al. (1960) used submerged glass slides which were liable to collect bacteria from a number of depths as they were lowered and raised through the water column. The results of his work may have some relative significance, however, and in particular it is claimed that discrete water types at depths down to 4000 m could be identified by their bacterial content. Sorokin (1964b) used direct counts of bacteria collected in sterile water samplers as a method for determining the total biomass of bacteria in the water column. From studies in the central Pacific, Sorokin (1964b) found that the biomass of bacteria in the euphotic zone was between approximately 10 and 50 mg/m³ which represents only a few per cent of the total particulate carbon. Below the euphotic zone the number of bacteria decreased sharply to 2 or 3 thousand per ml, while below 1000 m the number of bacteria were generally less than 1000 per ml or a total biomass of <0·2 mg/m³. The general form of these results agrees with biomass estimates of total living material made by Holm–Hansen (1969a) using ATP analyses as an indication of the amount of living organic carbon; however the total amount of living carbon determined from ATP analyses was considerably greater than that found from direct bacterial counts alone. This indicates that bacterioplankton, both in the euphotic zone and in deeper water, are generally a small fraction of the total biomass of living material and only a few per cent of the total particulate organic carbon in the water column. Relative changes in the biomass of bacterioplankton are generally more important, therefore, as an indication of recycling processes than as a direct source of food. However the methodology of bacterial counting still leaves a lot to be desired and it is probably correct to assume that many of the detrital particles in the ocean have bacteria accumulated on their surfaces and that this material is not included in direct counts of bacterial cells and clumps.

The presence of bacteria in the deep oceans (below 3800 m) has been reviewed by ZoBell (1968) who concluded that bacteria were widely but unevenly distributed at all depths. Numbers ranged from nil to 10^6/ml, the most dense populations having been found in materials from the sea floor. In near-shore environments, sedimentary material may be very rich in bacteria and it is believed that bacteria on the sedimented particles are the chief source of food for some benthic animals (Newell, 1965; Seki *et al.*, 1968).

Pelagic populations of organisms in the uppermost surface of the sea are referred to under the general term of neuston. Zaitsev (1961) first drew attention to the importance of neuston in the marine environment and a review of the subject has recently been given by Hempel and Weikert (1972). These authors subdivide the depth distribution of the neuston community into the "euneuston" — organisms with maximum abundance in the surface where they stay night and day; "facultative neuston" — organisms which concentrate at the surface only during certain hours, mostly during darkness; "pseudoneuston" — the maximum concentrations of these organisms do not lie at the surface but at deeper layers; however, the range of their vertical distribution reaches the surface layer at least during certain hours. The largest change in the population of neuston organisms occurs during the evening and at night when many organisms among the facultative neuston and pseudoneuston join the comparatively few euneuston species. Hempel and Weikert emphasize that while the population density of neuston organisms may be very high, the total biomass of neuston in the water column is small since only a very thin layer is occupied by this community.

1.3.4 TEMPORAL CHANGES IN PLANKTON COMMUNITIES

The most rapid temporal changes in a plankton community can be observed by continuous monitoring at a fixed point; these changes are not due to changes within the plankton community *per se* but are caused by internal waves. Armstrong and LaFond (1966) studied changes in nutrients, transparency, and temperature at a fixed point in a highly stratified near-shore environment; continuous 3 h records showed correlated fluctuations as illustrated in Fig. 16. The changes shown in Fig. 16 correspond to internal waves up to 5 m high with periods of around 10 min. Since the water intake used to collect these data was located at 9 m near the principal thermocline, the data reflect maximum changes due to an abrupt vertical gradient (e.g. from 0 to 12 μg at NO_3^-/l between 6 and 12 m). Superimposed on the short time scale changes in Fig. 16 are other oscillations caused by tide and alternating wind speed and direction.

Temporal changes within a plankton community itself are largely determined by the growth, mortality, sinking, and migration rates of the individual plankters and their pre-

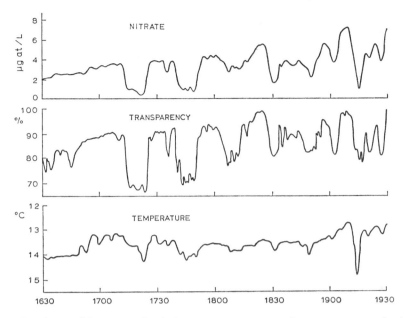

FIG. 16. Continuous 3-hour records of nitrate, transparency, and temperature at a fixed point (9·2 m from the sea floor) in a highly stratified environment off S. California (redrawn from Armstrong and LaFond, 1966).

dators. The most rapid growth or reproduction rates for phytoplankton are of the order of several hours but whole populations generally require at least a day or more to double in size. Bacterioplankton may generate within a matter of hours, depending on temperature and substrate concentration. Zooplankton growth rates vary enormously from less than a week for some protozoa to two years for antarctic euphausiids. Large populations of temperate copepods having one generation per year normally grow from egg to adult in 2 or 3 months, depending on food supply and temperature. Average mortality rates are generally less than growth rates or species would become readily extinct; however, over short periods, predation by planktivorous animals may cause mortality to exceed growth in a population. Sinking rates are discussed in Section 1.3.5; these, together with the rate of animal migrations may range from less than a meter to several thousand meters per day. Consequently observed temporal changes in a plankton community may vary depending on the frequency of an investigator's observations.

Changes which occur in a plankton community regularly every 24 hours are referred to as 'diel' changes; daily and nightly occurrences are called 'diurnal' and 'nocturnal', respectively. Included among diel changes are animal migrations, changes in photosynthetic potential, and, inshore, changes in plankton communities associated with the tidal cycle.

A detailed account of changes in an animal community caused by diel migrations is given by Bary (1967) as an example of a number of studies in this field. Using a 12 kHz sounder, Bary (1967) has described four stages in the ascent of organisms from a deep scattering layer (DSL) in a coastal environment. These stages started with a gradual vertical spreading of the DSL, one or two hours before sunset, followed by a period of slow ascent which started just before and ended after sunset. A period of rapid ascent followed about 1 h after sunset; during this period the migrating organisms came up at speeds ranging from *ca* 1 to 8 m/min. The final stage in the ascent migration was characterized by a reduced rate of ascent as the animals approached the surface, followed by a period when the animals gradually dispersed themselves in the surface layers. The descent followed the reverse process and commenced one or two hours before sunrise.

Diel variations in the rate of photosynthesis are quite apparent in that photosynthetic organisms require light for autotrophic growth; however, less obvious diurnal changes occur in the physiological response of phytoplankters to light. Shimada (1958) showed, for example, that photosynthesis of a plankton community reached a maximum during the early morning and declined during the rest of the day. This was apparently caused by a decrease in the amount of chlorophyll *a* during the latter part of the day. Steemann Nielsen and Jørgensen (1962) attributed this daily chlorophyll *a* rhythm to the fact that under laboratory conditions, no chlorophyll *a* was synthesized during the latter part of the day and consequently if herbivore grazing remained constant, a net decrease in chlorophyll *a* would be observed in nature. However, Sournia (1967) and others have shown that the photosynthetic index (mg C assimilated per mg Chl *a*) of phytoplankton is higher before noon than after noon. More recent data reported by Malone (1971) has indicated that changes in the photosynthetic index are complex and related to at least three factors including the time of day, the size of cells, and the availability of nutrients. Thus the photosynthetic index of microphytoplankton in tropical waters was highest after noon while the same index for nanoplankton was highest before noon. However in eutrophic waters nanoplankton showed their highest photosynthetic index after noon. These differences were not attributable to changes in cellular chlorophyll *a*. Thus apart from grazing effects, there is an obvious physiological change in photosynthesis which may effectively slow down any potential increase in phytoplankton standing stock during different parts of the day (see also Section 3.1.4).

Tidal changes in near-shore communities cause very marked fluctuations in the relative abundance of plankton and nekton. This is particularly apparent in estuarine communities where there is a large change in the type of water at a fixed point over a single tidal cycle. Welch and Isaac (1967) showed that chlorophyll *a* values could vary up to 800% in an estuary during 24 hours.

Large-scale temporal variations are associated with seasonal cycles in oceanic and neritic environments. A summary of seasonal cycles in plankton communities has been prepared by Heinrich (1962) and is represented in Fig. 17 with some modifications as suggested by the work of Sournia (1969). Four different seasonal cycles may generally be recognized in oceanic environments based on changes in standing stock of phytoplankton and zooplankton. The first of these (Fig. 17.1) is characteristic of arctic or antarctic waters where the amount of light is only sufficient for a single plankton bloom during the summer. The second seasonal cycle is characteristic of North Atlantic temperate waters (Fig. 17.2) where breeding of zooplankton cannot start until an increase in primary productivity has occurred in the spring. Thus as a result of the increased primary production, there is a temporary increase in the standing stock of phytoplankton followed by a decrease, as the grazing pressure from an increased standing stock of zooplankton becomes effective. In these areas there are generally two maxima in phytoplankton and zooplankton standing stocks, a large one occuring in the spring and a smaller one in the autumn. Theoretical discussions of factors causing an increase in plankton during the spring are given in Section 3.1.4; the occurrence of a second plankton maximum in the autumn is associated with similar factors including the amount of nutrient mixed into the water column after summer stabilization, the presence of sufficient light to cause photosynthesis and the occurrence of zooplankton species capable of taking advantage of an autumn increase in the standing stock of phytoplankton.

A third type of seasonal increase in plankton (Fig. 17.3) is found in the north Pacific ocean. In this environment neither the beginning of the zooplankton breeding nor the size of the zooplankton standing stock is dependent on the presence or abundance of phytoplankton in the early spring. Nauplii of the zooplankton species, *Calanus plumchrus* and *Calanus cristatus*, which predominate in this area are hatched from adults which have

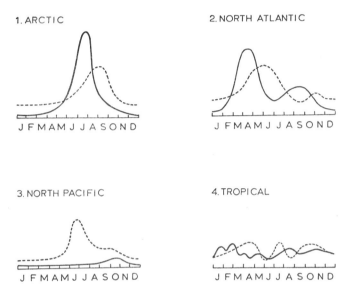

Fig. 17. Summary of seasonal cycles in plankton communities (—— changes in phytoplankton biomass; – – – changes in zooplankton biomass; modified from Heinrich, 1962).

wintered in deep water (>200 m) and laid their eggs without feeding in the spring. Consequently the young stages of these animals can take immediate advantage of any increase in primary productivity. In such environments it is difficult to observe any change in the phytoplankton standing stock (Fig. 17.3) except possibly during the autumn when a relaxation in zooplankton grazing allows a small increase in phytoplankton standing stock before winter.

A fourth type of seasonal cycle (Fig. 17.4) is found in tropical oceans. According to Sournia (1969) and Blackburn *et al.* (1970) there is very little evidence for predominant maxima and minima associated with seasonal events. A succession of small increases and decreases in phytoplankton and zooplankton standing stocks may occur throughout the year and these are largely determined by local weather conditions and the movement of water masses. Thus where a condition of upwelling occurs, such as in the Pacific approximately at the equator, a relatively high standing stock of plankton can be observed which is largely determined by the exact location of the equatorial counter current (Blackburn *et al.*, 1970). However, Owen and Zeitzschel (1970) observed a statistically significant seasonal change in primary production in the Eastern tropical Pacific; apparently this change was not reflected in changes in the plankton standing stock as discussed above. While the seasonal cycles in Fig. 17 indicate clear differences between different oceanic environments, it is also apparent that there are large zones of transition in the oceans. In sub-tropical waters at 32°N, for example, Menzel and Ryther (1960) observed a definite seasonal cycle in primary production in spite of an almost constant euphotic zone depth of 100 m.

Annual differences in the quantity of plankton measured at a fixed location may also be due to the strength of currents. This has been illustrated by Wickett (1967), who showed that annual variations in the concentration of zooplankton in the surface layers off California varied directly with the southerly transport of water during the previous year.

Conditions governing seasonal changes in neritic environments are much more complex than in oceanic environments due to the added effects of local geography, river discharge, and tides. Barnes (1956) described the seasonal occurrence of phytoplankton blooms and development of meroplanktonic barnacle larvae in the Firth of Clyde. From a study extending over ten years the author concluded that the normal spring increase in phytoplankton could be heavily suppressed in some years by strong winds which resulted in a sequential catastrophic reduction in the development of barnacle larvae. In the vicinity of large rivers the neritic oceanographic climate may be generally modified to provide nutrient entrainment and a more stable water column resulting in a higher standing stock of plankton. This is apparent, for example, in data presented by Anderson (1964) on seasonal changes in chlorophyll a off the Washington and Oregon coasts. In this study it is shown that chlorophyll a was higher in an area influenced by fresh water from the Columbia River (the Columbia River plume) than in either adjacent oceanic or neritic areas. Ice cover, and the effect of melting ice on the stability of the water column, may also cause earlier seasonal changes in plankton abundance in polar regions than in temperate oceanic waters located at lower latitudes (Marshall, 1958; Bunt and Lee, 1970).

Studies on the temporal variation in bacterial numbers and species have been reviewed by Sieburth (1968) using specific examples from Narragansett Bay, Rhode Island. The author showed that there were apparent effects of phytoplankton species, solar radiation and temperature on the number and type of bacteria present. For example, two different colonies of flavobacteria were present during the year, a yellow-pigmented colony being present during periods of low radiation and an orange-pigmented colony being present in larger numbers during periods of high solar radiation. Also inverse relationships between genera were found to occur independently of season. Thus pseudomonads appeared to be dependent on phytoplankton blooms and a percentage increase in the isolates of pseudomonads was accompanied by a percentage decrease in the isolates of arthrobacters. While these relationships appear as real temporal variations, it should be emphasized that their cause may be more complex than indicated by seasonal factors or by the presence of other organisms.

1.3.5 SOME PROCESSES GOVERNING PATCHINESS OF PLANKTON DISTRIBUTIONS

Stavn (1971) has summarized the principal factors determining the non-random or patchy distributions of planktonic organisms as follows:

1. Physical/chemical boundary conditions including light, temperature, and salinity gradients.
2. Advective effects as in wind or water transport, including small-scale variations due to turbulence.
3. Reproduction rates within the population.
4. Social behaviour with populations of the same species.
5. Coactive factors determined by competition between species.

While these factors may not account for all forms of patchiness, the environmental parameters included in the first and second items, together with the resultant physiological responses of planktonic organisms, are probably the principal causative agents in both small and large scale plankton patchiness.

From studies on light, temperature, and salinity gradients, it is apparent that planktonic organisms react strongly to various boundary conditions. For example, in experimental studies on salinity gradients Lance (1962) showed that the migration of the copepod *Acartia bifilosa* was inhibited when the bottom salinity ($34^0/_{00}$) was diluted at the top to more than 20 to $24^0/_{00}$. More sensitive changes were found by Harder (1968) among a wider group of planktonic organisms, some of which responded to salinity changes of less than $1^0/_{00}$. The light responses of plankton in the deep scattering layer has already been described (Section 1.3.3) but experimental studies, particularly with the freshwater cladoceran, *Daphnia*, have yielded some additional properties of the zooplankton/light response. Smith and Baylor (1953) showed that *Daphnia* was sensitive to different coloured light and that light of over 500 nm caused upward swimming while light of less than 500 nm caused downward swimming. McNaught and Hasler (1964) found an approximate linear relationship between the rate of vertical movement and the rate of change in the logarithm of the light intensity. Stavn (1971) showed that there was a minimum light intensity of 70 ergs/cm²/sec below which *Daphnia* did not respond to water movement; the author concluded that orientation in moving water was affected through visual detection of currents.

The distribution of plankton under various conditions of water movement is in part a passive response. Thus particles which tend to float will gather at a boundary where there is a convergence resulting in downwelling; conversely particles which tend to sink will accumulate at a divergence where there is upwelling (Stommel, 1949). Both convergences and divergences can be induced by wind and currents. In the case of a layer of warm water being pushed over a body of colder water, a convergence will result at the boundary of the two water masses. Conversely, a current which divides on meeting a headland may cause a divergence or upwelling. Plankton patchiness due to small-scale effects of wind was first observed by Nees (1949), who discovered that plankton samples collected parallel with the wind direction were more variable than samples collected at right angles to the wind. The reason for this is that under conditions of light wind, plankton accumulate in rows parallel to the wind (a property which can be observed visually in the Sargasso Sea, where *Sargassum* weed floats on the surface forming familiar wind rows). Thus if plankton are collected across the direction of the wind, the samples result in an integrated collection but if the samples are collected parallel to the wind, the resulting samples may come from a column of water in which plankton are concentrated, or from the relatively barren water in between the wind rows. The exact explanation for plankton patchiness under conditions of light wind may include several factors but in general it appears to be related to Langmuir circulation (Sutcliffe *et al.*, 1963; Stavn, 1971). The principal property of this circulation is that the water tends to move in vortices which result in "micro-zones" of upwelling and downwelling water (Fig. 18 and Faller, 1971, for a discussion of the physics of Langmuir circulation). Conditions for particle accumulations A, B, and C are based only on the buoyancy of particles as suggested by Stommel (1949); these conditions may apply particularly to phytoplankton and detrital accumulations or to all planktonic organisms in the absence of light. Under conditions, D, E, and F, however, experimental results obtained by Stavn (1971) are included; these take into account the relative current velocities of the water and the swimming speed of the animals together with their light reaction which (in the case of *Daphnia*) may be used to sense the currents. Thus during the day planktonic animals may swim into a

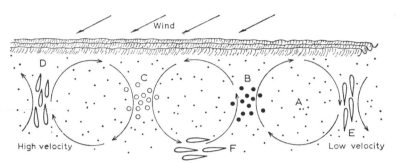

Fig. 18. Langmuir vortices and plankton distributions redrawn from Stavn (1971). A— neutrally buoyant particles randomly distributed. B— particles tending to sink, aggregated in upwellings. C— particles tending to float, aggregated in downwellings. D— organisms aggregated in high velocity upwelling, swimming down. E— organisms aggregated in low velocity downwelling, swimming up. F— organisms aggregated between downwellings and upwellings where there is less relative current velocity than within the vortices.

current (low velocity) since this is the most stable swimming position. In this case the animals will be aggregated in downwelling water and swimming up. At intermediate current velocities the plankton will be clumped in between the spirals and swimming horizontally against the current. At high current velocities animals may be trapped in a high velocity upwelling where they are attempting to swim down in order to avoid surface light which is generally too bright for many zooplankters (negative phototaxis).

Reproduction as a factor in causing plankton patchiness can be seen from examples in both the animal and plant communities. Figure 32 shows the development of a phytoplankton bloom which results in an uneven depth distribution of phytoplankton due to differences in the growth rate with light intensity. Uneven zooplankton distributions can result from the tendency of some zooplankters to release clumps of eggs; this contagious distribution of eggs may be carried through in the developing stages of copepods. Social behaviour among planktonic animals is probably only of minor significance in causing patchiness but of much greater significance among the nekton and in benthic communities. Coactive factors determined by competition between species may be caused by subtle effects, such as the release of substances toxic to other species, or more directly by predation patterns caused by different trophic levels in the plankton community.

Large scale patchiness, such as is depicted in Fig. 10 for phytoplankton production in the hydrosphere, can be mostly accounted for in terms of ocean circulation. LaFond and LaFond (1971) have discussed the patterns of water movement which lead to areas of high and low plankton productivity. The general mechanism involves some process which carries nutrient-rich deep water to the surface where there is adequate light for photosynthesis. Of particular importance in this respect is the influence of a land mass on circulation. LaFond and LaFond (1971) recognized three types of land influences on water movement which result in an upwelling of deep water, islands which create large eddy currents on the leeward side of the principal current flow, land promontories which cause similar eddies, and changes in underwater topography which cause turbulence in the flow of near surface currents.

In addition to these physical land mass effects, boundary conditions between currents moving in opposite directions may lead to large scale upwelling. This is particularly apparent along the equator in the Pacific Ocean where the equatorial current in the northern and southern hemispheres tends to move in a northerly and southerly direction, respectively (due to the Coriolis force, see Pickard, 1964, Fig. 27). This results in a divergence near the equator and an upwelling of deep water. Similarly a commonly occuring seasonal upwelling may be produced by winds blowing parallel to a coastline which can result in water being displaced offshore (an upwelling, or divergence) or onshore (a downwelling, or convergence). These observations are in accordance with physical oceanographic theory and are illustrated schematically in Fig. 19. According to Wooster and Reid (1963) the principal areas in the hydrosphere where an exchange occurs between near-surface and deeper waters

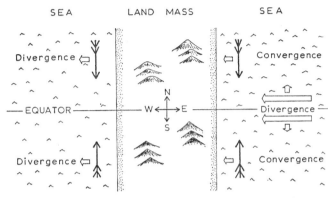

FIG. 19. Schematic effect of wind (⟶) and water (⇒) movement in producing areas of upwelling (divergence) and downwelling (convergence). Wind direction shown from the south toward the equator in the southern hemisphere and from the north toward the equator in the northern hemisphere. Water movement caused by this wind pattern results in a divergence off the west coast and a convergence off the east coast. Water movement is reversed with a reversal in wind direction. Currents along the equator in the northern and southern hemispheres similarly tend to move water towards or away from their boundary area, depending on the principal direction of water flow.

are (1) in high latitudes, (2) along the equator and, (3) in coastal regions, particularly on the eastern sides of oceans. This is apparent in the higher biological productivity of waters in such areas as the California current, the Peru current, and the Benguala current.

The vertical component of plankton patchiness may be partially influenced by factors discussed above, but in particular the processes of passive sinking and active vertical migration will influence plankton concentration at any depth.

Sinking rates of phytoplankton have been the subject of an extensive review by Smayda (1970a). From a theoretical approach, the rate of sinking (v) can be expressed as:

$$v = \frac{2gr^2}{9} \frac{(\varrho' - \varrho)}{\eta \cdot \phi_r}, \tag{14}$$

where r equals the radius of a particle, g is the acceleration due to gravity, ϱ' and ϱ represent the densities of the organism and the liquid, respectively, η is the viscosity of the liquid and ϕ_r is the "coefficient of form resistance". The latter term is given by

$$\phi_r = \frac{V_s}{V_a}, \tag{15}$$

where V_s is the terminal velocity of a sphere of volume and density equivalent to a non-spheroid body having a terminal velocity of V_a. From the above equations it is apparent that sinking rate will increase with diameter and density, but decrease with departures from a non-spherical shape (i.e. increase in ϕ_r); the subject is discussed by Munk and Riley (1952). For particles of *ca* 5 μ, the latter term is lowest for organisms having shapes approximating a plate, cylinder and sphere in increasing order; at about 50 μ plates and cylinders sink at about the same rate, but slower than spheres of the same volume, while for particles greater than 500 μ, spheres sink faster than cylinders but slower than plates. The sinking rates of plankton and other particles have been studied by a number of authors and a summary of their results has been made by Smayda and is represented in Table 2. From these results it is apparent that dead phytoplankton sink much faster than living phytoplankton and that larger sized organisms sink faster than smaller, as predicted by the above equation. However,

TABLE 2. A COMPARISON OF PHYTOPLANKTON SINKING RATES (AS m DAY^{-1}) WITH SOME RATES OBTAINED FOR DEAD MARINE PROTOZOANS, FISH EGGS, ZOOPLANKTON, FAECAL PELLETS, AND VIABLE MACROSCOPIC ALGAE, AS REPORTED ON IN THE LITERATURE FROM SMAYDA (1970a).

Group	Sinking rate (m day^{-1})	No. of species used	Reference
Phytoplankton			
Living	0–30	∼25	
Palmelloid stage	∼5–6150	1	As compiled in Appendix
Dead, intact	<1–510	10	Table I in Smayda, 1970a.
Fragments	$1 \cdot 5 – 26 \times 10^3$	?	
Protozoans			
Foraminifera	30–4800	—	
Radiolarians	∼350	—	p. 253 in Kuenen, 1950
Zooplankton			
Amphipoda	∼875	1	Apstein, 1910
Chaetognatha	∼435	1	Apstein, 1910
Cladocera	∼120–160	2	Apstein, 1910
Copepoda	36–720	14	Apstein, 1910; Gardiner, 1933;
Ostracoda	400	1	Apstein, 1910
Pteropoda	760–2270	5	Apstein, 1910;
Salpa	165–253	2	Apstein, 1910
Siphonophora	240	1	Apstein, 1910
'Animal plankton'	∼225–500	?	Seiwell and Seiwell, 1938
Faecal pellets	36–376	—	Osterberg *et al.*, 1963; Smayda, 1969.
Fish eggs	215–400	2	Apstein, 1910

it is difficult to generalize on the data in Table 2 since within the groups reported there are large differences in size and ash content (e.g. silica frustules among the phytoplankton) which cause a wide range in the sinking rates for any one group.

Mechanisms which tend to modify the sinking rate of phytoplankton include apparent changes in cell buoyancy and, among flagellates, the ability to swim in response to stimuli (e.g. light). Since some small flagellates can be observed under the microscope to swim a distance of their body length in less than a second, maximal swimming speeds for some species may be around 100 to 1000 μ per second, or in the approximate range of 1 to 10 m per day. Loeblich (1966) measured the swimming speed of *Gonyaulax polyedra* and from his observations, as well as others which he quotes, it appears that large dinoflagellates may swim at speeds from *ca* 2 to 20 m per day.

Changes in the buoyancy of phytoplankton cells may be accomplished through a change in the cellular constituents. Thus an increase in lipids or a change in the ionic content of the cell will alter the cell density. Of these two mechanisms, Smayda (1970a) considers that the latter mechanism is the more important, especially in cells having large vacuoles. The author reports on the work of Beklemishev, Petzikova, and Semina who analysed cell sap in the diatom, *Ethmodiscus rex*, and found that the total ion content of living cells was 23·9 mg/ml compared with 33·3 mg/ml in dead cells and 33·8 mg/ml in the seawater medium. The selective exclusion of heavy ions from cell sap as a means of buoyancy was first proposed by Gross and Zeuthen (1948) but it is apparent from observations made by Eppley *et al.* (1967) and theoretical calculations made by Smayda (1970a) that the theory does not entirely account for the range of density changes observed among the phytoplankton. Other physiological mechanisms yet to be fully explored are changes in buoyancy due to nutrient enrichment (Steele and Yentsch, 1960) and the effect of light (Smayda and Boleyn, 1966).

The vertical migration of zooplankton and their aggregation at specific depths has been reviewed by Banse (1964). In general vertical movement appears to be related either to the life history of a species or to its feeding habits. The former generally cause seasonal differences in the vertical distribution of plankton while the latter cause diel changes in vertical abundance. Vinogradov (1955) postulated in the latter case that food material was transported to great depths by a series of overlapping animal migrations. Thus it was considered that some bathypelagic animals obtained their food by migrating into the mesopelagic zone while some mesopelagic animals migrated into the epipelagic zone; the exact depth ranges being dependent on the species.

CHAPTER 2

CHEMICAL COMPOSITION

2.1 SEA WATER

The following section introduces a few important facts concerning the chemistry of sea water in relation to biological productivity. Since the chemistry of sea water has been treated in a number of comprehensive texts (e.g. Harvey, 1957; Riley and Chester, 1971) only a brief summary has been made here to cover some essential features of sea-water chemistry as they pertain to biological oceanographic processes. In addition, methods for determining some biologically important chemical constituents of sea water have already been given in previous texts (e.g. Barnes, 1959; Strickland and Parsons, 1968).

TABLE 3. CHEMICAL COMPOSITION OF SALT IN SEA WATER.
(CHLORINITY; 19·00‰).
(AFTER LYMAN AND FLEMING, 1940.)

Ion	%	Ion	%
Na^+	30·61	Cl^-	55·04
Mg^{2+}	3·69	SO_4^{2-}	7·68
Ca^{2+}	1·16	HCO_3^-	0·41
Sr^{2+}	0·03	Br^-	0·19
K^+	1·10	H_3BO_3	0·07

Oceanic water contains about 35 g of salt per kilogram of sea water; the major chemical components of the salt are shown in Table 3. In addition to these ions, sea water contains many minor elements and radicals, some of which have great biological importance in spite of their relatively low concentration. Elements which do not significantly change in concentration due to biological or geochemical activity (e.g. see Table 3) are referred to as being among the 'conservative' properties of sea water, while the 'non-conservative' substances include all materials whose concentrations are affected by biological or geochemical events. This distinction can be applied to short time periods but is not wholly correct for geological periods of time since the conservative compounds are in fact maintained at their apparent concentrations by an equilibrium which removes an amount equivalent to that being added by river runoff and other processes. A more exact distinction can be made in defining a 'conservative' constituent as one whose chemical activity is so low relative to the rate of physical oceanographic processes (such as mixing and advection) that its distribution is essentially controlled by physical oceanographic processes. Thus a non-conservative element is one whose distribution reflects the effects of short term biological or geochemical activity as well as the effects of physical oceanographic processes. Among the latter group are the nutrient salts, particularly nitrate and phosphate which may be present in deep ocean waters at concentrations of *ca* 30 and 2·5 µg at/l, respectively. The concentrations of these salts vary with the water mass, season, and depth. In many tropical and subtropical environments as well as in temperate waters following a phytoplankton bloom, the concentration of nitrates and phosphates may be below the limit of detection by conventional methods; in such cases it may be possible to demonstrate that the principal source of nitrogen is present in low concentrations (<5 µg at/l) of ammonia, urea, or some organic nitrogen compounds. Thus changes in the nutrients of sea water are usually reflected in a redistribution of biolog-

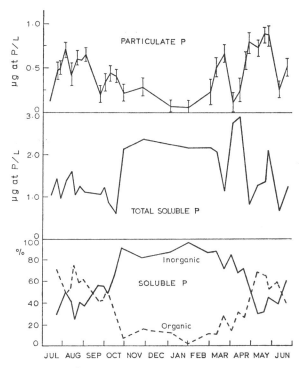

FIG. 20. Annual variations in the presence of three forms of phosphorus in a coastal environment, Departure Bay, British Columbia (redrawn from Strickland and Austin, 1960).

ically important elements; this is illustrated in Fig. 20, which shows changes in the concentration of three forms of phosphorus in a coastal environment over a period of one year. Biologically important constituents of sea water also include silicate, iron, manganese, cobalt, molybdenum, and other trace elements used in cellular metabolism. Carbon dioxide is present in sea water, mostly as bicarbonate in concentrations of *ca* 25 mg C/l; this level is a large excess over the amount generally required for plant growth and consequently the pH of sea water, which is largely determined by the bicarbonate/borate concentration, is usually in the range 7·5 to 8·5. Changes in pH within this approximate range are due to photosynthesis and respiration of marine organisms and low pH values are reflected in high concentrations of total carbon dioxide. Since carbon dioxide and oxygen are the products of biologically opposite processes, it is also usual to find high concentrations of oxygen (*ca* 4 to 8 ml/l) and higher pH values in areas of high photosynthesis, and lower oxygen concentrations (<4 ml/l) where respiratory processes predominate.

Dissolved organic material is present in open ocean sea water at a concentration of *ca* 0·4 to 2 mg C/l while particulate organic material, including both plankton and detritus (see Section 2.4) is present in concentrations ranging from *ca* 10 μg C/l in deep water to surface values of *ca* 100 to 500 μg C/l. The total concentration of organic material may not be as biologically important as the presence of certain organic constituents. Vitamin B_{12}, for example, may occur in sea water at concentrations of μμg/ml* and still be utilized by a number of organisms which require this vitamin for growth.

The most biologically important physical mechanism for altering the chemistry of sea water is the process of mixing. Near the sea surface, waters tend to become stratified, either by thermal heating or by differences in total salt content. In the tropical oceans, for example, heating causes stratification which effectively prevents deep nutrient-rich sea water from reaching the surface throughout the year. A similar situation can occur during the summer in temperate waters and nutrient exhaustion through plant growth in the surface layers results in low productivity in the absence of some mechanism for breaking down the stratified water column. In temperate latitudes, winter storms cause mixing of the water column so that deep water nutrients are seasonally restored to the surface layers. In other circumstances, wind and current-generated upwelling (such as off the coast of Peru) may cause an area of high productivity to occur during most of the year. Rivers entering the sea also cause mixing to occur and river water will entrain many times its own volume of deep salt water as it enters the sea; this, together with a supply of organic nutrients from the land, generally alters the chemistry of sea water in estuarine environments so as to cause local areas of very high productivity. Elementary discussions of physical processes governing mixing and stability of water masses have been given by Sverdrup *et al.* (1946) and Pickard (1964).

The nature of chemical changes caused by the admixture of river water with sea water to some extent depends upon the specific chemistry of local river waters. In general, however, both inorganic and organic constituents of sea water will be altered in the immediate vicinity of a river. Silicate is usually higher in river water than sea water and certain trace elements, such as thorium and cerium, may be several orders of magnitude more concentrated in river water (Goldberg, 1971). The natural heavy metal concentration of river waters has been difficult to determine in view of the widespread industrial use of metals such as copper, lead, and mercury. A sharp gradient in heavy metal concentration appears to exist in coastal waters, particularly in areas where there is some accountable source of heavy metal pollution (Abdullah *et al.*, 1972). The total dissolved organic constituents of large rivers entering the sea are generally higher than the surrounding sea water. While the types of organic compounds contributed by rivers are diverse and poorly investigated, particular biological interest has been placed on the vitamin content (e.g. Vitamin B_{12}; Burkholder and Burkholder, 1956) and the presence of chelating compounds. The latter may be closely associated with the humic acid content of river water; aside from their chelating properties, Prakash (1971) has indicated that humic acids may also enhance plant growth through certain specific physiological and biochemical reactions.

* 1 μμg. $\equiv$ 1 picog. $\equiv 10^{-3}$ nanog. $\equiv 10^{-6}$ microg. $\equiv 10^{-9}$ mg. $\equiv 10^{-12}$ g.

2.2 PHYTOPLANKTON

Table 4 shows the percentages of protein, carbohydrate, and fat found in different species of phytoplankton grown under the same conditions. Identification of major metabolites for the species reported was $100 \pm 10\%$ of the total organic matter. Considering the diversity

TABLE 4. PHYTOPLANKTON COMPOSITION OF MAJOR METABOLITES (FROM PARSONS *et al.*, 1961).

Species	Approx. cell volume (μ^3)	Percentage composition, ash free dry wt.		
		Protein[1]	Carbo-hydrate[2]	Fat[3]
Prasinophyceae				
Tetraselmis maculata	310	68	20	4
Chlorophyceae				
Dunaliella salina	400	58	32	7
Bacillariophyceae				
Skeletonema costatum	1390	58	33	7
Chrysophyceae				
Monochrysis lutheri	28	53	34	13
Dinophyceae				
Amphidinium carteri	740	36	39	23
Exuviaella sp.	780	35	42	17
Myxophyceae				
Agmenellum				
quadruplicatum	1·5	44	38	16

[1]Nitrogen $\times 6 \cdot 25$
[2]Anthrone reaction
[3]Saponifiable fraction only.

of size and taxonomic groups involved, the cellular composition of these species appears to be remarkably similar and shows, in contrast to terrestrial plants, a general predominance of protein over other constituents.

The relationship between cell size and total organic content of phytoplankton has been studied extensively by Strathmann (1967 and references cited therein), who gives two regression equations for the carbon content of different cell volumes. The first of these for phytoplankton, other than diatoms, is given as:

$$\log C = 0 \cdot 866 \log V - 0 \cdot 460. \tag{16}$$

Since diatoms contain a vacuole, the quantity of carbon per unit volume is less than for other phytoplankton species and is given as:

$$\log C = 0 \cdot 758 \log V - 0 \cdot 422, \tag{17}$$

where C is the carbon per cell in picograms and V is the cell volume in cubic microns.

From acid hydrolysates of different phytoplankton species, Cowey and Corner (1966) showed that most of the organic nitrogen could be accounted for in terms of amino acid nitrogen and that in confirmation of earlier work, the amino acid spectrum of different algae was very similar, as illustrated in Table 5. By comparison with protein of known nutritional value (casein), algal proteins show some differences in the proportion of amino acids but in

TABLE 5. AMINO ACID COMPOSITION OF UNICELLULAR ALGAE; WHOLE CELL HYDROLYSATES EXPRESSED AS g AMINO ACID N/100 g TOTAL N (FROM COWEY AND CORNER, 1966).

Amino acid	Organism					
	Skeletonema costatum	Phaeo-dactylum tricornutum	Monochrysis lutheri	Cricosphaera elongata	Chlorella ellipsoida	Whole[1] casein
Aspartic acid	7·92	7·05	6·13	6·52	5·7	7·1
Threonine	4·22	3·71	3·59	3·23	3·6	4·9
Serine	5·63	4·55	4·29	3·23	4·4	6·3
Glutamic acid	7·15	7·99	6·32	5·75	7·9	22·4
Proline	4·05	4·71	3·18	3·23	3·3	10·6
Glycine	8·89	6·69	6·72	5·61	7·9	2·0
Alanine	7·03	7·08	8·25	5·74	6·2	3·2
Valine	5·25	5·75	4·96	3·99	3·8	7·2
Methionine	0·71	1·29	1·59	2·27	0·8	2·8
Iso-leucine	4·14	3·33	2·91	2·04	2·4	6·1
Leucine	5·94	5·70	6·67	5·47	5·1	9·2
Tyrosine	1·51	1·80	2·19	2·55	2·1	6·3
Phenylalanine	3·22	3·03	2·79	2·95	3·0	5·0
Lysine	7·93	8·32	8·16	5·75	7·1	8·2
Histidine	3·08	3·28	3·55	3·29	3·2	3·1
Arginine	10·67	10·79	11·34	10·62	14·4	4·1
Tryptophan	—	—	1·74	0·98	2·1	1·7
Cystine/2	—	1·12	1·07	—	0·8	0·3
Total	87·34	86·19	91·94	78·53	89·9	110·5

[1]from Gordon and Whittier, 1966: (—) not reported.

general there is a well-balanced distribution which includes all the essential amino acids. Feeding experiments with rats, using the dinoflagellate, Gonyaulax polyedra, have further confirmed the dietary adequacy of phytoplankton protein (Patton et al., 1967). In a very extensive examination of the amino acid spectra in 31 species of marine phytoplankton, Chau et al. (1967) and Chuecas and Riley (1969) agreed with the general distribution of amino acids found by earlier workers but noted also a number of additional characteristics. These included the occurrence of amino acid derivatives of butyric and adipic acids, the frequent occurrence of serine as the principal amino acid in diatoms, and differences between diatoms and other algae in the relative proportions of a number of other amino acids.

The principal lipids found in three species of marine diatoms are shown in Table 6 (Lee et al., 1971). The particular characteristics of these data are the high concentration of phospholipid and the absence of wax esters. The latter are often the principal lipids in copepods which may feed upon diatoms. Fatty acids are included in the triglyceride, free fatty acid, and phospholipid fractions in Table 6 and their distribution in 8 classes of phyto-

TABLE 6. LIPID COMPOSITION OF DIATOMS FROM LEE *et al.* (1971).

	Lauderia borealis	*Skeletonema costatum*	*Chaetoceros curvisetus*
	(% recovery by weight)		
Hydrocarbon	11	16	2
Wax ester	0	0	0
Triglyceride	16	14	12
Sterol	17	15	10
Free fatty acid	5	12	16
Phospholipid	51	57	50
Total % dry wt.	13·2	8·6	9·1

plankton is shown in Fig. 21 (from Ackman *et al.*, 1968). Early data on the fatty acid composition of plankton are difficult to interpret since most of the long chain fatty acids were missed in analyses. However, in general marine phytoplankton fatty acids contain between 10 and 30% palmitic acid (C_{16}) as well as appreciable amounts of C_{18}, C_{20}, and C_{22} polyunsaturated acids*. Among the Bacillariophyceae the major saturated acids are C_{14} and C_{16} and the unsaturated acids are 16 : 1, 16 : 3 and 20 : 5. As an exception to other classes of algae, C_{18} acids appear to be very minor constituents of the Bacillariophyceae. The

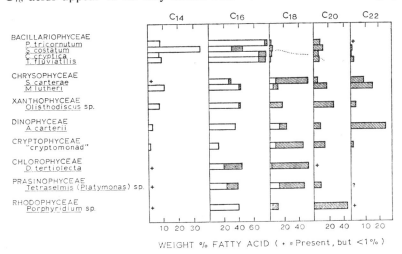

FIG. 21. Fatty acid spectrum of 12 marine phytoplankters. ▢ saturated or one double bond; ▨ polyunsaturated acids (redrawn from Ackman et al., 1968).

Chrysophyceae differ from the Bacillariophyceae in containing more polyunsaturated acids, particularly C_{18} and C_{22}. The Dinophyceae appear to be characterized by 16 : 0 and polyunsaturated C_{20} and C_{22} acids. However, Chuecas and Riley (1969) showed that two species of Dinophyceae contained large amounts of 16 : 0 fatty acid, while in a third the principal fatty acid was 16 : 1. Four members of the Cryptophyceae examined by Chuecas

* C_{18} indicates a chain length of 18 C atoms; 16 : 1 indicates a chain length of 16 C atoms and 1 double bond; 20 : 5 ω 3 indicates 3 carbon atoms between the terminal methyl group and the middle of the double bond nearest the terminal methyl group.

and Riley (1969) were similar and the principal acids were 16 : 0, 16 : 1, 18 : 3, 18 : 4, 20 : 1 and 20 : 5. These results differ from the results found by Ackman *et al.* (1968), who reported a virtual absence of 16 : 1 and 20 : 1 acids while Chuecas and Riley (1969) described a relatively high concentration of the latter as a characteristic of the Cryptophyceae. The representatives of the Chlorophyceae and Prasinophyceae shown in Fig. 21 have very similar compositions; the principal components were 16 : 0 and polyunsaturated C_{16} and C_{18} acids. In *Porphyridium* sp. the principal fatty acids were palmitic acid and C_{20} polyunsaturated acids including the unusual occurrence of over 20% of 20 : 4. The taxonomic position of the Xanthophyceae, *Olisthodiscus* sp. is uncertain, but its fatty acids composition was similar to the two Chrysophyceae. Parker *et al.* (1967) investigated the fatty acid composition of 11 representatives of the Cyanophyceae and found a predominance of 16 : 0, 16 : 1, 18 : 1 and 18 : 2 acids; however, there were obvious differences between species including one species in which half of the fatty acids were C_{10}. Branch-chained fatty acids were absent from the Cyanophyceae but were a major component in marine bacteria.

The carbohydrates of marine phytoplankton occur either as storage products of cellular metabolism or as constituents of cell wall material; small quantities of other carbohydrates may be associated with cellular metabolic processes. Using *Skeletonema costatum*, Handa

TABLE 7. CARBOHYDRATE FRACTIONS OBTAINED FROM
Skeletonema costatum (FROM HANDA, 1969).

Carbohydrate fraction	Yield (%)	Principal sugar
Water—extractable	45·6	91% glucose
Soluble in ethanol–acetone	(10·8)	Glucose and oligosaccharides
Insoluble in ethanol–acetone	(34·8)	β-1, 3-glucan
Residue	54·4	55% mannose

(1969) demonstrated that approximately half of the algal carbohydrate was water soluble (Table 7) and that this consisted primarily of a glucose polymer, β-1, 3-glucan; the water insoluble residue consisted of mannose with lesser amounts of rhamnose, fucose, and xylose. The carbohydrate storage product of diatoms is very similar to laminarin, the reserve polysaccharide of the brown seaweeds. The diatom polysaccharide is known as chrysolaminarin which differs from laminarin in that although it is a glucose polymer it contains no mannitol which is a constituent of laminarin. Members of the Chrysophyceae may also contain chrysolaminarin (Beattie *et al.*, 1961, and references cited therein). The dinoflagellate, *Thecadinium inclinatum*, has been reported to contain an α-1, 4-glucan (Vogel and Meeuse, 1968); water soluble extracts of the Rhodophyceae may also contain an α-glucan, floridean starch.

The carbohydrate constituents of cell wall material are complex but are usually known to include homopolysaccharides, such as cellulose, mannan, and xylan, or some heteropolysaccharides, such as glucuromannan found in the Bacillariophyceae (Ford and Percival, 1965). From these observations Handa and Tominaga (1969) concluded that the cell wall carbohydrate, represented by the residue in Table 7, was not a single compound but that the diverse monosaccharide constituents found in acid hydrolysates represented sugars from several polysaccharides.

Monosaccharides obtained from acid hydrolysates of whole cells are shown in Table 8 for algal species harvested during the exponential phase of growth. The amount of 'crude fibre' found in different species may be interpreted as indicating the proportion of the total carbohydrate employed as cell wall material. Glucose, galactose, and ribose were found in all species analysed; glucose was always the predominant sugar which agrees with more recent results for 3 species of Bacillariophyceae analysed by Handa and Yanagi (1969). However, these authors found approximately 30% of the monosaccharides to be mannose which is a

TABLE 8. MONOSACCHARIDE COMPOSITION OF WHOLE HYDROLYSATES OF UNICELLULAR ALGAE (FROM PARSONS *et al.*, 1961).

	Crude fibre (percentage of total carbohydrate)	Principal sugars (percentage dry weight of cells)										
		Glucose	Galactose	Mannose	Ribose	Xylose	Arabinose	Rhamnose	Fucose	Fructose	Hexosamine	Hexuronic acids
CHLOROPHYCEAE												
Dunaliella salina	9·8	17·2	11·8	—	1·7	—	—	—	—	—	—	+
PRASINOPHYCEAE												
Tetraselmis maculata	12·6	11·9	2·3	—	0·95	—	—	—	—	—	—	+
CHRYSOPHYCEAE												
Monochrysis lutheri	3·6	22·1	4·4	—	1·3	3·5	—	—	—	—	—	+
HAPTOPHYCEAE												
Syracosphaera carterae	1·7	9·2	7·1	—	1·5	0·8	1·9	—	—	—	—	+
BACILLARIOPHYCEAE												
Chaetoceros sp.	22·8	3·3	1·5	0·79	0·71	0·4	—	2·8	+	—	—	+
Skeletonema costatum	9·6	16·4	1·8	0·87	1·2	—	—	1·0	0·9	—	—	+
Coscinodiscus wailsii	29·0	2·1	0·4	0·41	+	—	—	0·7	0·5	—	—	+
Phaeodactylum tricornutum	2·5	10·7	2·7	3·7	0·72	0·7	—	1·5	—	—	—	+
DINOPHYCEAE												
Amphidinium carteri	2·0	19·0	8·4	—	0·9	—	—	+	—	—	—	—
Exuviaella sp.	37·0	26·8	8·3	—	+	+	+	+	—	—	—	—
MYXOPHYCEAE												
Agmenellum quadruplicatum	17·4	17·4	3·2	—	1·5	—	—	—	—	3·5	0·3	+

+Sugars detected but not estimated.
−Sugars not detected.

considerably larger fraction than is shown for the Bacillariophyceae in Table 8. Further differences between the results in Table 8 and other authors may also occur. For example, Lewin *et al.* (1958) reported the presence of fucose in *Phaeodactylum tricornutum*. A note of caution is required in accepting data presented in Table 8 since these analyses were performed using paper chromatography and better methods are now available.

Major metabolites and some minor cellular constituents in 5 species of unicellular algae have been reported by Ricketts (1966a). Among minor cellular constituents, nucleic acids ranged from 1 to 7%; DNA 0·31 to 0·86%; RNA 0·7 to 6·65%; phospholipids 1·0 to 4·0%; acid soluble phosphorus 0·13 to 0·69%; phosphoprotein phosphorus 0·02 to 0·14%; total phosphorus 0·49 to 1·19% of the dry weight. Total phosphorus has been analysed by a number of authors and may range from about 0·5 to 3% of the dry weight depending to some extent on the supply of phosphate in the surrounding water. Holm–Hansen (1969b) measured the DNA content of 10 species of marine phytoplankton and found that the amount ranged from 0·01 to 200 pg per cell. This range was strongly correlated with the amount of carbon per cell and the ratio of DNA to cell carbon was 1 : 100.

The ash content of some phytoplankton species may be very high due mainly to cell wall constituents, such as calcium carbonate in coccolithophores and silica in diatoms. Silicon is widely distributed in small amounts throughout all plants and is generally present in phytoplankton at levels between 0·1 and 1% dry weight; among diatoms the silica frustule may represent up to *ca* 40% of the dry weight of the cells but this can vary by a factor of at least 5 depending on the availability of silicon in the surrounding water (Lewin, 1957; Vinogradov, 1953; Parsons *et al.*, 1961). In a series of papers Coombs *et al.* (1967a, b and c) have shown that silicon uptake and deposition in the cell wall of the diatom *Navicula pelliculosa* required ATP and that metabolic changes accompanied silicon starvation. These included a decrease in the net synthesis of proteins, carbohydrates, chlorophyll, and fucoxanthin, and an increase in the net synthesis of diadinoxanthin and lipid. Among coccolithophores, the calcium carbonate coccoliths originate from within the cell (Paasche, 1962) and may form a dense cover over the cell wall or at other times they may be totally absent (e.g. Braarud, 1963). The distribution of 18 trace metals in 15 species of phytoplankton has been studied by Riley and Roth (1971) and compared with earlier work by Vinogradova and

TABLE 9. DISTRIBUTION OF TRACE ELEMENTS IN PHYTOPLANKTON
(FROM RILEY AND ROTH, 1971).

	Chlorella[1] *salina*	*Asterionella*[1] *japonica*	*Phaeo-dactylum*[1] *tricornutum*	Sea[2] plankton
% ash, dry wt.	9·5	24·1	7·9	45·8
Element (ppm)				
Mn	48	54	73	118
Zn	301	115	325	282
Cu	25	105	110	36
Ag	4·6	10	6·6	3·3
Pb	<10	<20	46·3	900
Sn	<9	35	101	34
Ni	3·1	<12	6·2	48
Be	2·7	<4	3·8	<6
V	3·7	<5	<2	3·1
Al	118	1750	490	<5000
Ti	13·5	85	16	940
Ba	70·5	75	95	248
Cr	<3	5·5	4·4	7·5
Sr	31·4	<7	7·6	70

[1]Grown on synthetic medium.
[2]Sample from the Irish Sea.

Koval'skiy (1962); some of the former analyses are reported in Table 9. The principal conclusion reached from analysing species from the same class was that trace element distribution was not correlated with taxonomy and further that the concentration of individual elements may vary considerably between species. From further experiments Riley and Roth (1971) were able to show that in general the trace metal content of algae could be increased by increasing the concentration of metals in the medium in which the organisms were grown.

The photosynthetic pigments of unicellular algae are probably their most exhaustively analysed components. Early work has been summarized by Strain (1951) and Goodwin (1955); later constritions include reports by Goodwin (1957), Strain (1958), Haxo and Fork (1959), 'OhEocha and Raftery (1959), Allen *et al.* (1960 and 1964), Dales (1960),

TABLE 10. PIGMENTS OF MARINE PHYTOPLANKTON.

CLASS

Pigment	Bacillario-phyceae	Dino-phyceae	Chryso-phyceae	Chloro-phyceae	Myxo-phyceae	Xantho-phyceae	Crypto-phyceae	Prasino-phyceae	Hapto-phyceae
Chlorophyll a	+++	+++	+++	+++	+++	+++	+++	+++	+++
b				++				++	
c	++	++	+			(+)	++		++
Carotene α							+++	+	
β	+++	+++	+++	+++	+++	+++		+++	+++
γ								+	
Xanthophylls									
Fucoxanthin	+++	(+)	+++			+++			+++
Neofucoxanthin	++		++						++
Diadinoxanthin	++	++	++			++			++
Diatoxanthin	+					+			+
Dinoxanthin		+							
Peridinin		+++							
Neoperidinin		+							
Lutein				+++				++	
Zeaxanthin				+				+	
Flavoxanthin				+					
Violaxanthin			+	+				++	
Neoxanthin				+					
Alloxanthin (1+2)							+++		
Monodoxanthin							+		
Crocoxanthin							++		
Myxoxanthin					++				
Myxoxanthophyll					++				
Antharaxanthin					(+)				
Siphonaxanthin				+*				+	
Number of unidentified pigments		2		1		1	2	8	1
Phycobilins					++		++		

( ) Presence uncertain, +++ Principal pigment, ++ Generally reported as present, + Sometimes reported as present.
*Only in the siphonous (non-planktonic) members.

Jeffrey (1961), Parsons (1961), Jeffrey and Allen (1964), Ricketts (1967a and b, 1970), Riley and Wilson (1967) and Riley and Segar (1969). Some of the more recent reports have included extensive quantitative data. However, apart from the use of quantitative data on chlorophyll *a* and total carotenoids in ecological studies, the chief importance of pigment data lies in the obvious differences in pigment composition of different taxonomic groups. A summary of these data is presented in Table 10, which shows that all marine algae contain chlorophyll *a* and some accessory pigments, including chlorophylls and carotenoids. Secondary chlorophylls appear to be absent in the Myxophyceae and possibly in the Xanthophyceae. The tentative presence of chlorophyll *c* in the latter is based on its presence in *Olisthodiscus*, which may have been wrongly placed in the Xanthophyceae (Riley and Wilson, 1967). Chlorophyll *c* occurs in the Bacillariophyceae, Dinophyceae, Chrysophyceae, Cryptophyceae, and Haptophyceae; chlorophyll *b* occurs in the Chlorophyceae and the Prasinophyceae. Beta carotene is the predominant carotenoid in all the unicellular algae except in the Cryptophyceae where α-carotene predominates. The Prasinophyceae appear to be unique in containing several carotenes and possibly lycopene (Ricketts, 1970). Fucoxanthin is usually the predominant xanthophyll in the Bacillariophyceae, Chrysophyceae, Xanthophyceae, and Haptophyceae while peridinin characterizes the Dinophyceae. Other xanthophylls are more or less characteristic of the taxonomic classes. However, the Prasinophyceae contain a large number of xanthophylls and many of these have not been properly identified (e.g. see Ricketts, 1970). In addition, inconsistencies occur; for example, Riley and Wilson (1967) have reported on the occurrence of fucoxanthin as the principal xanthophyll in *Gymnodinium veneficum*—peridinin was absent from this species; astaxanthin, which is usually a characteristic pigment in some marine animals, was found by Jeffrey (1961) to occur in a Chlorophyceae; antheraxanthin was found by Parsons (1961) to be the principal xanthophyll in a Myxophyceae in which both myxoxanthin and myxoxanthophyll were apparently absent. The reporting of chlorophyll *c* as a single compound in Table 10 may be open to question since Jeffrey (1969) has identified two spectrally different components in chlorophyll *c* preparations. In general, however, it remains the conviction of pigment chemists (e.g. Strain, 1966) that there is a relationship between the occurrence of particular pigment systems and the taxonomic classification of an organism. In this respect it has been suggested (Ricketts, 1966b) that the very diverse spectrum of pigments found in the Prasinophyceae may reflect their possible position at a branching point in protistan evolution.

From a consideration of photosynthetic mechanisms the diversity of pigments discussed above can be divided into three major pigment systems. These are known as

1. the chlorophyll *a* and *b* system,
2. the chlorophyll *a*, *c* and carotenoid system,
3. the chlorophyll *a* and phycobilin system.

Photosynthetic processes in the Chlorophyceae, Bacillariophyceae, and Myxophyceae are characteristic of the three systems, respectively.

The chemical composition of unicellular algae can be greatly affected by changes in environmental conditions; one of the earliest illustrations of this was with *Chlorella* (Spoehr and Milner, 1949). Subsequent investigations have been carried out by a number of authors and some differences between the results of different analysts may be attributed to a lack of clear definitions of the environmental conditions under which the algae were grown. Environmental conditions which may affect composition include nutrient levels, light, temperature, and salinity. An illustration of changes in the composition of a phytoplankton bloom is given in Table 11. The results are taken from Antia *et al.* (1963). In the experiments reported,

TABLE 11. CHANGES IN THE CHEMISTRY OF SEA WATER AND PHYTOPLANKTON DURING AN ALGAL BLOOM*
(FROM ANTIA et al., 1963).

Day	Nitrate	Phosphate (μg at/l)	Silicate	Phytoplankton Ratios (by weight)								
				$\frac{Protein}{Carbon}$	$\frac{Carbohydrate}{Carbon}$	$\frac{Lipid}{Carbon}$	$\frac{C}{N}$	$\frac{C}{P}$	$\frac{Si}{C}$	$\frac{C}{Chl\ a}$	$\frac{Carotenoids}{Chl\ a}$	$\frac{N}{P}$ (atoms)
12	17	1·5	48	2·1	0·2	0·1	3	17	0·5	37	1·0	13
14	9	0·9	42	2·0	0·3	0·2	3	22	0·6	23	0·9	16
16	0	0·2	15	1·9	0·6	0·3	3	27	1·0	25	0·9	18
18	0·5	0·2	10	1·6	1·1	0·3	4	31	0·9	37	1·0	18
20	0·5	0·3	2	1·5	1·0	0·4	4	37	0·9	49	1·0	19
24	1·0	0·4	1	1·4	1·2	0·3	5	33	—	52	1·1	16
27	0·5	0·3	2	1·3	1·1	0·4	5	32	—	66	1·2	15
30	0	0·5	4	1·3	0·9	0·3	5	35	—	79	1·2	17
Vigorously growing phytoplankton with excess nitrate in the water				2·0	0·2	0·15	3	20	0·6	25	0·95	12–15
Unhealthy phytoplankton in nitrate depleted water				1·3	1·1	0·35	4·5	33	1·0	60	1·15	15–18
Values suggested by Strickland (1960)				—	—	—	6 ±2	40 ±15	0·8	30	0·9	14·5

*Predominantly diatom.

a large volume of sea water was enclosed *in situ* within a plastic sphere and changes in the sea water chemistry and phytoplankton were followed over a period of 30 days. The principal phytoplankton species which grew during the experiment were a mixture of diatoms (*Thalassiosira rotula, T. aestivalis, Skeletonema costatum, Stephanopyxis turris, Chaetoceros pelagicus, Navicula* sp.) and a dinoflagellate (*Gyrodinium spirale*). From the results of the experiment it may be seen that the protein content of the phytoplankton decreased as nitrate became exhausted; carbohydrate and lipid increased as protein decreased. A summary of the overall changes in cell components, together with a summary of values obtained from the literature (Strickland, 1960) are presented in the bottom half of Table 11. These ratios are useful in ecological studies involving trophic relationships where it is often impossible to make separate measurements of plant and detrital particulate material *in situ*.

Studies carried out by Handa (1969) on chemical changes in a pure culture of *Skeletonema costatum* showed that the transfer of healthy cells from light to dark resulted in a 57% decrease in carbohydrate followed by a 28% decrease in protein and a 44% decrease in lipid. The principal glucose constituent utilized was the water soluble fraction (β-1, 3-glucan) with little change occurring in the quantity of cell wall constituents. Ackman *et al.* (1968) studied time-dependent changes in the fatty acid content of 3 species of phytoplankton. Indefinite results were obtained with one species but *Dunaliella tertiolecta* showed a decrease in C_{16} unsaturated acids and an increase in C_{18} unsaturated acids over a two-week period. The effect of temperature on fatty acid synthesis in *Monochrysis lutheri* was also investigated; total polyunsaturated acids were found in twice the quantity at 10 °C compared with 20 °C cultures, and 18 : 1 ω 7 was reduced at lower temperatures from 2·5 to 0·8%. There is an apparent parallel in these results and results which have been obtained with terrestrial plants (Ackman *et al.*, 1968). Degens (1970) studied changes in protein and amino acids of three species of unicellular algae grown at different temperatures in nutrient enriched water and following respiration in the dark for up to 19 days. The author concluded that there was a gain in protein content with increasing water temperature but that the total protein content decreased with the length of respiration. The conclusion reached by Degens (1970) that in the initial stages of respiration, protein is lost more rapidly than carbohydrate, is not born out by Handa's results with respect to the rapid utilization of the water-soluble carbohydrate fraction during respiration.

The effect of light quality on the chemical composition of phytoplankton has been studied extensively by Wallen and Geen, both in the laboratory (1971a and b) and under natural conditions (1971c). From laboratory studies it was apparent that the colour of light influenced the pathway of $^{14}CO_2$ metabolism; for example, blue or green light favoured protein synthesis while white light favoured carbohydrate synthesis. Minor constituents, such as DNA, RNA, and pigment concentrations were also influenced by the colour of light. These results were largely borne out by field studies. For example, the relative activity of ^{14}C in the phytoplankton carbohydrate fraction decreased with depth while the ^{14}C incorporated into the protein fraction increased with depth. From these results the authors concluded that depth differences in light quality were independent of changes in light intensity in determining the chemical composition of phytoplankton *in situ*.

It has been suggested (Blumer *et al.*, 1971, and references cited therein) that differences in the hydrocarbon content of different species of marine phytoplankton are sufficiently great to be used as a means of identifying algal classes and possibly species. This suggestion may be particularly useful in determining the composition of mixed phytoplankton populations. In addition, since hydrocarbons are not readily destroyed in passing up the food chain, it appears that a useful technique may be available for diagnosing food preferences of zooplankton and higher organisms, *in situ*. The predominant hydrocarbon found in the Bacillariophyceae, Dinophyceae, Cryptophyceae, Haptophyceae, and Euglenophyceae was the unsaturated n-21 : 6 heneicosahexaene (abbr. HEH). This compound was absent in the Cyanophyceae, Rhodophyceae, Xanthophyceae, and Chlorophyceae, which were generally characterized by n-14 to n-17 hydrocarbons and in particular by either n-pentadecane or n-heptadecane. The hydrocarbons of phytoplankton appear to be very different from the hydrocarbons of zooplankton (mostly C_{19} and C_{20} isoprenoid alkanes and alkenes) and of mineral oils, in which olefins are absent but which contain branched chain, alicyclic, and aromatic compounds in relatively high proportions.

2.3 ZOOPLANKTON

Major chemical constituents of the zooplankton from three areas are shown in Table 12. While some disagreement between similar groups may be due to differences in analytical technique (especially drying), it is apparent that the data show general similarities. Thus the dry to wet weight ratio of copepods and euphausiids appears to be in the range 10–20% with possibly higher values found among amphipods; high dry weights in pteropods may

TABLE 12. MAJOR CHEMICAL CONSTITUENTS OF ZOOPLANKTON.

Organism or group	Dry wt. as % wet wt.	Carbon	Nitrogen	Hydrogen	Phosphorus	Ash	Comment
		(all constituents expressed as a % of dry wt.)					
Copepods*	11·6–16·3	35–48	8·2–11·2		0·7–0·8		Sargassa Sea plankton from
Euphausiids & mysids*	14·5–18·0	35–43	9·4–10·5		1·4–1·6		Beers (1966)—
Chaetognaths*	6·0– 7·4	22–34	6·3– 9·4		0·5–0·7		range for each group
Fish/fish larvae*	11·9–16·0	33–42	8·3–10·7		0·9–1·8		
Polychaetes*	5·7–27·0	16–44	4·4–11·2		0·4–1·8		
Siphonophores*	0·3– 6·1	3–16	1·0– 4·4		<0·1–0·2		
Hydromedusae*	0·3–10·1	5–10	1·4– 6·9		0·1–0·4		
Pteropods*	22–32	21–25	2·7– 4·2		0·2–0·4		
Copepods+	10·2–15·8	32–42	4·7– 7·1		0·4–0·8	18–23	Continental
Euphausiids+	19·0–20·0	33–37	5·2– 7·1		0·9–1·2	19–22	shelf off New
Ctenophores+	4·7– 5·0	(6·4)	0·2– 1·1		0·1–0·2	70–75	York, from
Pteropods+	3·5–19	26–28	2·2– 5·0		0·3–0·6	24–64	Curl (1962)—
Tunicates+	4·0– 4·1	7–11	0·3– 1·5		0·1–·03	71–77	range for each group
Copepods*	9·2–33·9	39–66	5·1–13·1	6·7–10·3		2–6	North Pacific
Amphipods*	18·4–36·6	26–48	4·4– 8·2	4·4– 7·6		10–37	from Omori
Euphausiids*	20·2–21·3	39–47	10·0–10·7	6·7– 7·6		8–9	(1969)—range
Chaetognaths*	11·6–14·1	44–48	10·7–11·1	7·2– 7·6		4–5	for each group
Pteropods*	25·0–36·4	17–29	1·5– 6·0	1·1– 3·8		29–43	

*Dried at 60° C
+Dried at 105° C

be associated with inorganic shell material, which is generally reflected in an ash content of greater than 20%. On a dry weight basis, the ash content of 70% encountered among the ctenophores and tunicates is due to inorganic salts contained in the gelatinous bodies of these animals. The carbon content of a wide variety of zooplankton ranges from *ca* 30 to 40% of the dry weight, except when a large amount of ash is present; similarly nitrogen and phosphorus values generally lie in the range 5–10 and 0·5–1·0%, respectively. The C/N ratio for more than 80% of the samples analysed by Omori (1969) was in the range 3–8.

Platt *et al.* (1969) found a high degree of correlation ($r = 0.94$) between the carbon content and caloric equivalent of marine zooplankton. Using their own data and data from other reports, the best fitting equation was given as

$$y = -3370 + 136x - 0.514\,x^2, \tag{18}$$

where y is in cal/g dry wt. and x is the % organic matter in the dry plankton. If data on ash content were included a higher correlation ($r = 0.98$) was obtained. For the authors' area of study this was given as

$$\text{cal/g dry wt.} = 1351 + 106\,(\%\ \text{carbon}) - 21.2\,(\%\ \text{ash}). \tag{19}$$

Raymont *et al.* (1969a and references cited therein) have analysed a large number of zooplankton species (decapods, mysids, and euphausiids) for their biochemical constituents in terms of total protein, carbohydrate, and lipid. The authors concluded that in all their zooplankton analyses, protein was high (53–64% dry weight), carbohydrate was extremely low (1–3%), and lipid content was variable. Some areal and vertical differences were encountered; neritic mysids had a high protein (70–72%) and low lipid (13–14%) content, while some deep sea and offshore species had a lipid content of greater than 20%. Ash content was highest in deep sea decapods (15–24%) and lowest in neritic mysids (7–8%). The chitin content of all species of crustacea ranged from 3 to 8%. Differences reported above may in part be due to seasonal changes in composition for a single species. Thus Raymont *et al.* (1969a, b) showed that for the euphausiid, *Meganyctiphanes norvegica*, seasonal changes in lipid content varied from 10 to 30%, and in protein content from 50 to 60%; the seasonal relationship between protein and ash content was the reciprocal of the lipid content.

The amino acid constituents of two species of euphausiid (from the south and north Pacific) and one species of copepod from the English Channel are compared with the amino acids from casein in Table 13. According to Cowey and Corner (1963) the proportions of amino acids found in *Calanus helgolandicus* are very similar to those found in phytoplankton and particulate matter on which the animals feed. There are some differences between the proportion of amino acids in the two euphausiids compared with *Calanus helgolandicus*, particularly with reference to the amounts of glycine and alanine and the sulfur amino acids, cystine and methionine. However, in general the amino acid spectrum of the three crustaceans appears well balanced and comparable to a nutritionally reliable animal protein, casein.

Jeffries (1969) has studied the free amino acid pool in a temperate zooplankton community. From these studies it was shown that free amino acids were generally higher in summer than in winter and that major changes occurred during periods of community stress, such as those accompanying environmental changes during spring and early winter. Differences were reflected in quantitative changes in the concentration of free amino acids per unit weight of tissue, while relative amino acid composition was the same both in summer and winter zooplankton. The most abundant free amino acids were taurine, proline, glycine, alanine,

TABLE 13. AMINO ACID COMPOSITION OF ZOOPLANKTON.

Amino acid	Species (g amino acid/100 g protein)			Whole casein[3]
	Euphausia pacifica[1]	Euphausia superba[1]	Calanus helgolandicus[2]	
Alanine	5·61	5·46	8·1	3·2
Glycine	5·35	4·67	8·9	2·0
Valine	5·19	5·90	6·6	7·2
Leucine	7·83	7·70	7·8	9·2
Isoleucine	5·16	5·10	4·8	6·1
Proline	3·47	4·21	4·1	10·6
Phenylalanine	6·50	6·47	4·1	5·0
Tyrosine	4·15	4·06	1·5	6·3
Tryptophan	1·57	1·50	—	1·7
Serine	4·82	4·95	4·1	6·3
Threonine	4·83	4·70	4·0	4·9
Cystine/2	1·35	1·45	0·7	0·3
Methionine	3·25	3·03	1·2	2·8
Arginine	5·95	6·22	7·8	4·1
Histidine	2·22	2·30	1·8	3·1
Lysine	7·84	8·58	8·1	8·2
Aspartic acid	13·7	12·2	9·4	7·1
Glutamic acid	14·7	14·6	11·8	22·4
Glucosamine	2·04	3·45	—	—
Amine N	1·40	1·37	—	—
Taurine	—	—	2·1	—

[1]From Suyama et. al. (1965); [2]from Cowey and Corner (1963), and converted from % amino N to % amino acid; [3]from Gordon and Whittier (1966).

and arginine. The changes reported are consistent with metabolic changes in other animals where the free amino acid pattern is reported to represent a picture of an organism's metabolic activities; the author suggests that these changes could be used to monitor subtle environmental changes.

The major lipid fractions in two copepods are shown in Table 14. The most characteristic

TABLE 14. COPEPOD LIPIDS (FROM LEE et al., 1970).

Fraction	Calanus[1] helgolandicus	Gaussia[1] princeps
	(% recovery by weight)	
Hydrocarbons	3	Trace
Wax esters	30	73
Triglycerides	4	9
Polar lipids[2]	17	} 17
Phospholipids[3]	45	
Total lipid as % dry weight	15	28·9

[1]Collected off La Jolla, California.
[2]Free acids, cholesterol, mono- and diglycerides.
[3]Lecethin and phosphatidyl ethanolamine.

feature of this table is the presence of large amounts of wax esters, which are not generally found among the phytoplankton. Lee *et al.* (1971) found that the lipid content, and spectrum of fatty acids and alcohols, was largely dependent on the copepods diet. Thus there was a linear correlation between the amount of food fed and total lipid content of the zooplankton; further, the composition of wax esters was changed with increased diet. The most obvious feature of this change was an increase in the amount of C_{30} ester from 12 to 37% in well-fed animals to *ca* 50% or greater in animals on a minimal diet. The chain length and degree of saturation of fatty acids and alcohols of the wax esters are shown in Table 15, together

TABLE 15. *Calanus helgolandicus* LONG CHAIN ALCOHOL AND FATTY ACIDS (FROM LEE *et al.*, 1971)—% WEIGHT AS METHYL ESTERS.

Component	Wax esters		Phospholipid
	Alcohols	Fatty acids	
14 : 0	1·5	11·8	1·5
16 : 0	13·0	29·0	39·8
16 : 1	1·0	13·2	tr
16 : 2	—	—	0·2
16 : 3	—	tr	tr
17 : 1	1·2	0·6	—
18 : 0	0·8	4·3	4·1
18 : 1	3·8	12·1	2·7
18 : 2	0·2	3·6	0·4
20 : 1	18·4	7·3	0·2
20 : 2	tr	5·1	0·3
20 : 3	—	0·6	—
20 : 4	40·9	7·6	0·3
20 : 5	—	2·0	13·3
22 : 3	8·4	—	—
22 : 6	8·1	2·1	36·5
24 : 3 to 7	1·7	—	—

with the fatty acid components of the phospholipids. The authors found that the triglyceride and free fatty acids generally resembled the dietary fatty acids but that the long chain alcohols did not correspond either in chain length or degree of saturation with the dietary fatty acids. From this it was concluded that the alcohols were derived from dietary fats by a variety of different metabolic pathways. The fatty acids of the phospholipid fraction (Table 15) were dissimilar to the fatty acids of the wax esters as indicated primarily by the large amounts of 20 : 5 and 22 : 6 fatty acids found in the former. These fatty acids were not present in large quantities in the diet fed to the copepods and it is assumed that they must have been synthesized by the animals. The observation from feeding experiments that the phospholipid fatty acids were not affected by changes in the amount or type of food ingested is compatible with their structural function in animal metabolism; in contrast the wax esters appear to be entirely storage products of copepod metabolism. While these results have dealt almost exclusively with one species of copepod found off California, similar results regarding the high wax ester content of copepods has been found by Yamada and Ota (1970) with respect to *Calanus plumchrus* in the western subarctic Pacific Ocean. Two North Atlantic tunicate filter feeders, *Pyrosoma* and *Salpa cylindrica*, have been shown

(Culkin and Morris, 1970) to contain appreciable amounts of 14 : 0 (myristic), 16 : 0 (palmitic), 16 : 1 (palmitoleic), and 18 : 1 (oleic), as well as the polyunsaturated 20 : 5 and 22 : 6 acids which were also found in the phospholipid fraction of *Calanus helgolandicus*. Lewis (1969) has reported that minor fatty acids in the amphipod, *Apherusa glacialis*, included branched-chain acids and positional isomers; small amounts of branched-chain fatty acids were also found in the two tunicates by Culkin and Morris (1970), who suggested that they may be synthesized within the zooplankton. In a study of fatty acids in phytoplankton, zooplankton, and fish, Williams (1965) showed that the proportional transfer of fatty acids from phytoplankton to zooplankton, discussed above, was also maintained between zooplankton and fish. Thus throughout the three levels of production in the pelagic food chain there is a ubiquitous predominance of palmitic acid with generally large proportions of myristic, palmitoleic, and oleic acids in many organisms.

Jeffries (1970) has discussed the effect of seasonal changes in the environment on the fatty acid composition of zooplankton in Narragansett Bay, Rhode Island. The ratio of palmitoleic to palmitic declined during thermal warming from 2·0 to *ca* 0·3, which reflected a dietary change from diatoms to dinoflagellates. The two acids were found to vary reciprocally throughout the year, as were oleic and steraidonic acids. The author suggests that such variations in chemical patterns could be useful in studying biological organization at the community level.

2.4 DETRITUS

The subject of detritus in sea water has been discussed in a number of recent reviews (e.g. Nishizawa, 1969; Riley, 1970). Unlike plankton, detritus is difficult to define either from its appearance or through its chemical composition. However, while the word lacks definition it should not be regarded as lacking in importance.

Microscopic observations of particulate material from sea water collected on a membrane filter, or in a settling cylinder, always show the presence of a large number of particles of irregular shapes and sizes. These particles are collectively referred to as detritus although Odum and de la Cruz (1963) have pointed out that scientists use a wide variety of terms to describe particulate material in aquatic environments. Thus the terms 'organic debris', 'suspended matter', 'particulate organic and inorganic material', 'leptopel', and (in lakes) 'tripton' may be assumed to refer to detritus unless specifically defined in another sense (e.g. the word leptopel in geology is sometimes used to describe a fine mud or clay). The term 'seston' should be used in referring to all particulate material including living organisms (plankton) and detritus. Detritus may be further qualified as being inorganic detritus and organic detritus, or bio-detritus; the latter terms imply that the detritus has originated from dead organisms. However, detritus has micro-organisms associated with it and might be more appropriately thought of as microcosm consisting of a particulate substrate in which bacteria and other micro-organisms may be embedded. From this concept the term 'organic aggregate' has often been used as a better description of detritus seen under the microscope. Kane (1967) differentiated between two types of aggregates found in the Ligurian Sea; a 'typical' aggregate was composed of a substrate to which various recognizable particles, such as bacteria and phytoplankton, adhered. These aggregates were usually brownish-yellow and were generally between 30 and 50 μ in their longest dimension. 'Granular' aggregates were composed of small inorganic grey-black granules and were considered to be a possible early stage in the development of 'typical' aggregates. Nemoto and Ishikawa (1969) used various stains to identify the nature of detrital aggregates in the East China Sea. Acid fuchsin, which is a general stain for cytoplasm, was found to stain many particles at all depths; however, the ratio of particles stained with Millon's reagent (protein stain) to particles stained with acid fuchsin increased with depth. A few particles stained with α-naphthol (carbohydrates) and with Sudan black (fats). From similar studies carried out in the north Atlantic, Gordon (1970a) differentiated between detrital particles which appeared as aggregates, flakes and fragments; judging from reactions to histochemical stains he concluded that the aggregates were chiefly carbohydrate, the flakes were chiefly protein and the fragments were entirely carbohydrate.

There is a lack of agreement among analysts on the ratio of carbon to nitrogen in detrital material from below the euphotic zone. The C : N ratio for compounds such as urea is less than one while for most proteins it is about 4·5; the latter value may also characterize some bacteria (Porter, 1946). In phytoplankton the ratio of C : N is 3 to 6 (See Table 11) while sediments generally have a C : N ratio of 10 or more (e.g. Seki *et al.*, 1968; Degens, 1970). It would be logical to expect, therefore, that deep water detritus would have a C : N ratio between that of phytoplankton and sediments. Some authors (e.g. Holm–Hansen *et al.*, 1966; Handa, 1968; Gordon, 1970b) have obtained C : N ratios of deep water detritus of greater than 10; others (e.g. Parsons and Strickland, 1962a; Menzel and Ryther, 1964; Dal Pont and Newell, 1963) have obtained values of less than 5. Differences also exist in the C : N ratio of soluble organic material. For example Duursma (1960) found values in deep water of *ca* 3 while Holm–Hansen *et al.* (1966) have reported values of *ca* 10. While these values may be real, it must also be considered that differences in analytical procedures may have led to different results. For example if the C : N ratio of deep water soluble organic material is assumed to be low (*ca* 3) and if this fraction is partly included with the particulate material (e.g. through adsorption on $MgCO_3$-coated membrane filters, such as were used by Parsons and Strickland, 1962a) it would tend to decrease the apparent C : N ratio of the particulate material. Alternatively if bacteria form an appreciable fraction of the detrital biomass they will tend to cause a lower C : N ratio than if the detritus is derived entirely from phytoplankton. In experimental studies on the decomposition of phytoplankton in the absence of zooplankton, Otsuki and Hanya (1968) showed that the C : N ratio of dissolved material released by phytoplankton gradually decreased over a period of 200 days from *ca* 10 to <3. Decomposition of organic nitrogen in phytodetritus was rapid in the first 30 days; release of carbon and nitrogen during this period amounted to 65–70% of the total. Approximately 10% was converted to dissolved substances and the rest remineralized as CO_2 and NH_4^+.

Williams and Gordon (1970) studied the $^{13}C/^{12}C$ ratios (expressed relative to a standard) in dissolved and particulate organic material down to 4000 m in the Gulf of Mexico and off the coast of southern California. For deep water they found similar ratios ($-21·2$ to $-24·4$) regardless of location or season; further the ratio for dissolved organic material was very similar to the ratio for particulate organic material ($-22·0$ to $-24·3$). It was also shown that these ratios corresponded most closely to the cellulose and 'lignin' fraction (residue after all other extractions) of phytoplankton ($-22·4$ and $-23·1$, respectively). These ratios were also similar to the ratio for organic material from sediments ($-20·8$ to $-22·3$) but were quite different from organic material derived from the Amazon River ($-28·5$ to $-29·5$). From these detailed studies the authors concluded that deep-water organic detritus is derived primarily from marine plankton and that the soluble and particulate fractions are similar in chemical composition. In another report Williams *et al.* (1969) determined the age of deep water soluble organic materials being 3400 years old; this value agrees within an order of magnitude with values derived by Skopintsev (1966) from a theoretical approximation based on input and the rate of decomposition.

Various methods have been used to determine how much of the detrital particulate matter is biologically utilizable, particularly in the euphotic zone where it is associated with large populations of filter-feeding animals. The principal difficulty in this respect has been to separate or differentiate between the detrital particulate material and the living particulate material (mostly phytoplankton). In a study on the biological oxidation of organic detritus in which a correction was made for the amount of phytoplankton, Menzel and Goering (1966) found that in samples from 1 m in the north Atlantic, between 16 and 52% of the detritus was biodegradable; applying the same technique to samples taken from below the euphotic zone (200 to 1000 m) the authors could not detect any oxidation of organic material and concluded that the material was essentially all refractory. From feeding experiments, Paffenhöfer and Strickland (1970) found that *Calanus helgolandicus* would not feed directly off detritus. This is similar to the result obtained by Seki *et al.* (1968), who showed that *Artemia* would not feed directly off sedimented plant material but that they could feed off bacterial aggregates which were grown using the detritus as a substrate.

Gordon (1970b), using proteases, showed that approximately 20 to 25% of deep organic detritus was hydrolysable. Holm–Hansen and Booth (1966) determined the amount of ATP in particulate material at various depths off the coast of California. Since ATP is rapidly destroyed following the death of an organism, the total quantity of ATP in organic material can be used as an index of the amount of living material (from extensive studies on living organisms the ratio currently used by Dr. Holm–Hansen is $ATP \times 250 =$ cellular carbon). In deep water samples, 500 to 1000 m, the authors concluded that 3% or less of the particulate material was alive; above 100 m, 14 to 79% was living while at intermediate depths approximately 6% was living with one exception of 27%. Thus it appears that the deep water hydrolysable material found by Gordon (1970b) represented dead biodegradable detritus; this is in contrast with the results found by Menzel and Goering (1966) using a different technique.

Direct chemical analyses were made on particulate material taken from 400 m in the north Pacific (Parsons and Strickland, 1962a). From the discussion presented above, it may be concluded that most of the organic material in this fraction was dead. The analyses showed that the material was composed of protein and carbohydrate; the principal amino acids were glycine and alanine with glutamic acid, aspartic acid, lysine, arginine, serine, and proline also being detected. Degens (1970) has reported in greater detail on the amino acid composition of deep water detritus in the Atlantic. Below 200 m there was an apparent increase in the proportions of serine, glycine, lysine, and arginine, and a decrease in alanine with depth down to 2500 m.

The carbohydrate fraction of deep water detritus analysed by Parsons and Strickland (1962a) was 70% insoluble to treatment with weak acid and alkali (in contrast with 'crude fibre' values for healthy phytoplankton in Table 8). The principal sugars following total acid hydrolysis were glucose, galactose, mannose, arabinose, and xylose. The quantity of fat present was less than 1%. Glucosamine and hexuronic acids were not detected indicating the lack of appreciable amounts of chitin from crustaceans or hexuronides from marine plants, respectively. However, Wheeler (1967) has reported on the presence of copepod carcasses between 2000 and 4000 m in the north Atlantic; as chitinous material these would contribute to the total organic detritus, but on the basis of their concentration per m³, they would only account for between *ca* 0·5 and 5% of the total organic carbon in deep water. Handa and Tominaga (1969) and Handa and Yanagi (1969) have carried out detailed carbohydrate analyses of particulate materials down to 700 m in the northwest Pacific.

Their results show that the water soluble carbohydrate fraction of phytoplankton disappeared between 50 and 300 m and that between 300 and 1000 m, only water insoluble carbohydrates remained. Detrital material from these depths contained 50% less glucose than phytoplankton, but correspondingly higher proportions of galactose, mannose, xylose, and, in contrast to Parsons and Strickland (1962a), appreciable quantities of glucuronic acid.

The inorganic fraction of particulate detritus may be a variable fraction of the total dry weight but generally it amounts to at least 70% (Wangersky, 1965). The exact chemical nature of the inorganic material has not been defined but calcium carbonate particles are known to occur in the open ocean (Wangersky and Gordon, 1965) and these may be intimately associated with the organic material in sea water (Chave, 1965 and 1970). In coastal areas inorganic particles are associated with clays and other minerals derived from the land; these may include appreciable amounts of silicon, iron, aluminium, and calcium (Armstrong and Atkins, 1950). More detailed data on the inorganic fraction of detritus can be derived from literature on marine sediments (e.g. Griffin et al., 1968).

The quantity of chlorophyll a in particulate material decreases with depth so that below the euphotic zone most of the chlorophyll a has either disappeared or been converted to phaeophytin or phaeophorbide (Lorenzen, 1965). Saijo (1969) reported on chlorophyll pigments down to 4000 m in the northwest Pacific. The total concentration of chlorophyll a in waters below 400 m was from <0·001 to 0·003 μg/l; phaeo-pigments were often present, however, at ten times the concentration of chlorophyll a. Currie (1962) showed earlier that digestion of phytoplankton by zooplankton resulted in the almost total conversion of chlorophyll a to phaeo-pigments. From studies in the western Indian Ocean, Yentsch (1965) showed that the ratio of chlorophyll a to phaeopigments decreased with light intensity. This was concluded to be a reversible reaction since the chlorophyll to phaeophytin ratio of dark adapted phytoplankton could be increased when they were restored to the light. Carotenoids appear to be much more resistant to decomposition both within the water column (Yentsch and Ryther, 1959) or as a result of the digestive processes of animals. In the latter case, Fox et al. (1944) found an average content of 24 mg of carotenoid per 100 g dry weight of feces from mussels, but negligible quantities of chlorophyll.

Considerable biological importance has recently been attached to the surface film of organic material in the sea (Garrett, 1965; Harvey, 1966) Using a specially constructed surface skimmer, Harvey (1966) showed that the surface film contained large amounts of living nanoplankton, structural components of disintegrated organisms, surface active substances, chlorophyll, and carotenoid pigments. Garrett (1964, 1967) showed that major chemical components of this layer were fatty acids, fatty acid esters, alcohols and hydrocarbons. Taguchi and Nakajima (1971) measured the ratio of particulate carbon in the surface layer (150 μ thick) compared with subsurface (10–15 cm deep) samples. The ratio of surface to subsurface particulate carbon was generally 2 to 5 with a maximum concentration factor of 17·6 in coastal environments. Similar high ratios were found by Nishizawa (1971) in the open ocean waters of the equatorial Pacific. Both Sieburth (1971), and Taguchi and Nakajima (1971) have drawn attention to the concentrations of living organisms (neuston) associated with the detrital surface film. Bacterial counts were generally one or two orders of magnitude higher in the surface film than just below the surface; the species of phytoplankton and zooplankton were also quantitatively greater and to some extent qualitatively different, in the surface layers compared with subsurface samples. These observations have lead to the conclusion that a neuston community exists at the sea surface which constitutes a micro-environment rich in a specific microflora and fauna.

The presence of this neuston community may in part be due to an accumulation of material floating on the sea surface. However, it may also be formed by the peculiar properties of the air–water interface. Sutcliffe *et al.* (1963) showed that when air is bubbled through filtered sea water, organic particles are formed. It was further shown that these particles could support the growth of brine shrimp (Baylor and Sutcliffe, 1963). Carlucci and Williams (1965) showed that the action of bubbling sea water tended to concentrate bacteria at the surface. However Menzel (1966) and Barber (1966) have questioned some of the experimental evidence for these findings. In the latter reference it is reported that neither bacteria nor bubbling alone cause a significant increase in the amount of particulate material in sea water containing organic materials with a molecular weight of less than 100, 000.

While most observations on detritus have been made on samples collected in water bottles it is apparent that the form of detritus as observed under the microscope may bare little relation to its appearance *in situ*. From some of the earliest *in situ* observations made under the sea it was noted that much of the detritus was aggregated into clumps or long streaks which were easily disintegrated by the action of sampling bottles. Suzuki and Kato (1953) described its appearance *in situ* as "marine snow". Aggregation of particulate materials has been studied by a number of authors (e.g. Sheldon *et al.*, 1967; Parsons and Seki, 1970) and it appears that bacteria are probably involved in this process, either through certain species which tend to form clumps or through the release of organic polymers which tend to attach to other particles. The effect of aggregation is to increase the size and therefore the sinking rate of the particles. However as Riley (1970) and Kajihara (1971) have observed the sinking of detrital aggregates does not obey Stoke's law [eqn. (14)] since there is a decrease in the specific gravity of particles with increased size due to aggregation.

CHAPTER 3

THE PRIMARY FORMATION
OF PARTICULATE MATERIALS

3.1 AUTOTROPHIC PROCESSES

In the ocean there are algae and some bacteria which can synthesize high-energy organic compounds from low-energy inorganic compounds such as water and carbon dioxide. The source of energy for these organisms is either light, or chemical energy derived from the oxidation of inorganic compounds; such organisms do not require organic materials as a source of energy. This life style is called 'autotrophy' and the organisms are called 'autotrophs'. When one considers the cycling of organic material in the oceans, autotrophic organisms are referred to as 'primary producers' because it is these organisms which are the only producers of original autochthonous organic material in the sea. The organic material produced by the primary producers is referred to as 'primary production' and primary production per unit time in a unit volume of water (or under a unit of area) is called 'primary productivity'. On the basis of differences in energy source for organic matter synthesis, autotrophy is divided into two different categories known as 'photosynthesis' (light energy) and 'chemosynthesis' (chemical energy).

3.1.1 PHOTOSYNTHESIS

The fundamental relationship governing the photosynthetic process can be summarized in the following equation:

$$n\text{CO}_2 + 2n\text{H}_2\text{A} \xrightarrow{\text{light}} n(\text{CH}_2\text{O}) + 2n\text{A} + n\text{H}_2\text{O} \tag{20}$$

where reduced compounds, such as H_2O, H_2, H_2S, $\text{H}_2\text{S}_2\text{O}_3$, and some organic compounds may be used as the H-donor in H_2A but only light is used as the energy source.

Photo-autotrophs in the ocean include representatives of the algae as well as photosynthetic bacteria; both types of these organisms are usually widely distributed in the ocean. However, quantitatively the algae are the most important photo-autotrophs in the ocean, with a few exceptions to this generalization to be found in neritic regions.

Photosynthetic algae require H_2O as the H-donor, and equation (20) can be modified for algal photosynthesis as follows:

$$n\text{CO}_2 + 2n\text{H}_2\text{O} \xrightarrow{\text{light}} n(\text{CH}_2\text{O}) + n\text{O}_2 + n\text{H}_2\text{O}. \tag{21}$$

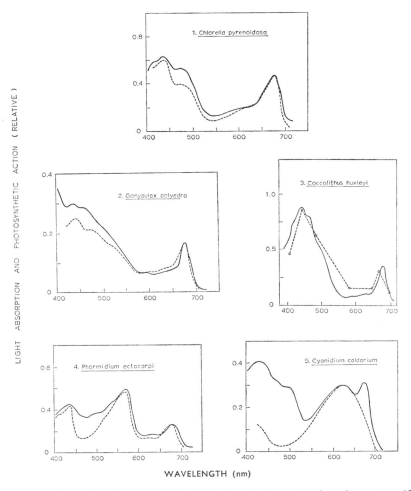

FIG. 22. Light absorbtion of intact cells (solid line) and photosynthetic action spectra (dotted line). 1, 2, 4, and 5 redrawn from Haxo (1960); 3, after Paasche (1966).

This process requires energy of *ca* 120 Kcal per mole of carbohydrate formed. The energy is derived through the absorption of light by chlorophyll *a* and accessory pigments contained in the chromatophores of the algal cells. The wavelength of light absorbed by photosynthetic pigments is mainly in the region of visible light from 400 to 700 nm, but light absorption patterns are different in each algal group, depending on their pigment systems (see Table 10). Fig. 22 shows light absorption spectra of some intact marine algal cells. Light of wavelengths shorter than 600 nm is mainly absorbed by chlorophyll *a* and accessory pigments. Above 600 nm light for photosynthetic processes is only absorbed by chlorophyll. The latter absorption peak is normally observed at 680 nm; however, by improved techniques, such as derivative spectrophotometry and spectrofluorometry at very low temperatures, the existence of two other peaks, one at 670 and the other at 695 nm, were recognized (Kok and Hoch, 1961). These multiple peaks of chlorophyll *a* are distinctive for intact cells and are not observed in chlorophyll *a* extracted in organic solvents. From these findings it became clear that chlorophyll *a* has a complex morphological structure within the chloroplast.

The function of these absorption peaks is not fully understood but two separate photo-chemical reactions are now recognized based on light absorption at the shorter wavelength (called chlorophyll *a* 670) and the longer wavelength (called chlorophyll *a* 680).

Energy absorbed at the longer wavelength (chlorophyll *a* 680) is used directly for photo-chemical reactions, but energy absorbed at the shorter wavelengths is transferred by the accessary pigments to chlorophyll *a* 670 before being used. The energy accepted by both

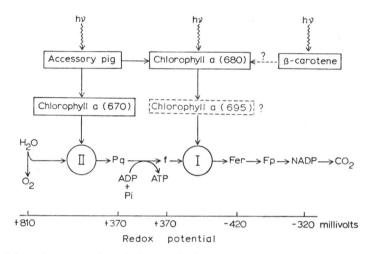

FIG. 23. Schematic presentation of photosynthetic system in algae, represented by two photo-chemical systems (I and II); Pq, plastoquinone; f, cytochrome; Fer, ferredoxin; Fp, NADP reductase; ADP and ATP, adenosine di- and tri- phosphates; Pi, inorganic phosphate; hν, light energy. Dotted lines indicate possible connections in the absence of clear experimental evidence (redrawn from Fujita, 1970).

types of chlorophyll *a* is used for photochemical reactions in two photosynthetic systems, I and II (Fig. 23). These two photosynthetic systems are conjugated by a series of electron transfers, involving quinone and cytochrome. System I, which is primarily mediated through energy derived from chlorophyll *a* 680, is mainly involved in electron transfers. The photo-chemical reaction catalysed by pigment System II liberates oxygen from water and transfers electrons to plastoquinone (Pq in Fig. 23). Energy transferred from the two photochemical reactions is used for (1) the reduction of nicotinamide adenine dinucleotide phosphate (NADP) and (2) photophosphorylation of adenosine diphosphate (ADP) into the high-energy compound, adenosine triphosphate (ATP). This series of reactions is carried out in the light and they are collectively referred to as the 'light reaction'. The reducing power of NADPH and the energy of ATP promote the reduction of CO_2 and produce carbohydrate as well as synthesizing proteins and fats. These reactions are carried out in the dark and are referred to collectively as the 'dark reaction'. Together the light and dark reactions are generally included in the term photosynthesis and this whole process takes place within the chloroplasts. The metabolic processes of the dark reaction were first demonstrated by Calvin and his colleagues, and thus it is also known as the 'Calvin cycle'. The details of the Calvin cycle can be found in the treatise of Calvin and Baasham (1962).

The efficiency of energy transfer from pigments to photosynthetic systems is not always the same. The transfer efficiency can be estimated from the 'quantum yield' which expresses how many moles of CO_2 are fixed by one photon of light absorbed by pigments. Some

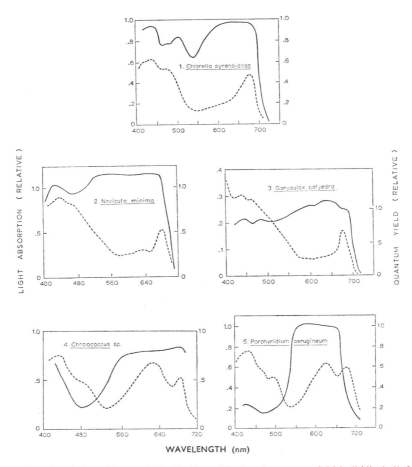

FIG. 24. Light absorbtion of intact algal cells (dotted line) and quantum yield (solid line). (1, 3, and 5 redrawn from Haxo, 1960; 2 redrawn from Tanada, 1951; 4 redrawn from Emerson and Lewis, 1942).

examples of quantum yields are shown in Fig. 24 for different marine algae. The transfer efficiency of light to chlorophyll a is highest because the energy is transferred directly to the photosynthetic system; light is transferred via the accessory pigments with variable efficiencies. For example, the accessory pigments of diatoms, dinoflagellates, and coccolithophorids transfer light energy with an efficiency similar to that of chlorophyll a, but the accessory pigments of green and blue-green algae have relatively low transfer efficiencies. From a consideration of the photosynthetic efficiency of pigments, Fujita (1970) classified marine algal groups into (1) chlorophyll a and b type for green and euglenoid algae, (2) chlorophyll a, c, and carotenoid type for diatoms, dinoflagellates, and brown algae, and (3) chlorophyll a and phycobilin type for red and blue-green algae. The actual light utilization spectra of algae can be obtained by combining the light absorption of intact cells with the quantum yield; such a curve is called the 'action spectrum' (see Fig. 22). The action spectrum is considered to show the photosynthetic light utilization efficiency of a cell and it is one of the important characteristics of a species since it determines the ability of phytoplankton to adapt to different light regimes in the ocean. The action spectrum is not stable, however, but sometimes changes depending on growth conditions such as differences in illumination, wavelength of light and nutrient concentration. Extreme examples of change are observed in the blue–green algae (Halldal, 1958; Fujita, 1970).

The final product of photosynthesis shown in eqn. (20) and (21) is carbohydrate, $n(CH_2O)$. This means that the photosynthetic quotient expressed as the ratio of evolved O_2 to absorbed CO_2 (PQ) is close to unity. The photosynthetic quotient becomes higher if other organic compounds are produced (e.g. ca 1·25 for proteins and 1·43 for lipids). The photosynthetic quotient is not a stable property of a cell but it changes depending on the past history of the species and environmental conditions (e.g. Myers, 1953, observed changes from 1·04 to 2·50 in $Chlorella$ $pyrenoidosa$).

Photosynthetic products are partly consumed by respiration, which is the reverse reaction of photosynthesis. Respiration takes place both in the light and in the dark; however, it can usually only be detected experimentally as O_2 consumption or CO_2 production in the absence of light (i.e. with no photosynthesis). Measurement of phytoplankton respiration in the ocean is practically impossible, firstly because of its very low activity and secondly because of the respiration of other micro-organisms (e.g. bacteria and zooplankton). Since healthy algal cells in culture have shown a respiration of about 10% of maximum gross photosynthesis (P_{max})—(Steemann Nielsen and Hansen, 1959b; McAllister et $al.$, 1964)— the indirect estimation of respiration from P_{max} is now used as a matter of convenience. However, such estimates require further investigation since it has been suggested that the respiration rate of algae (per unit biomass) may change depending on the surrounding environmental conditions (Ryther and Guillard, 1962; Bunt, 1965) and the physiological state of the algae (Nihei et $al.$, 1954). Also it has recently been shown that under certain conditions the rate of respiration of some algae increases markedly (2 to 3 times) in the light (Gregory, 1971). This enhancement of respiration in the light has been called "photorespiration" and the mechanism is not at present understood. When dinoflagellates are abundant in the water the ratio of respiration to P_{max} should be changed because high respiration rates of 35 to 60% P_{max} were observed in cultures of $Exuviaella$ $cordata$, $Gymnodinium$ $wulffii$, $Peridinium$ $trochoideum$, and $Prorocentrum$ $micans$ (Moshkina, 1961). These high respiration rates were attributed to the motility of dinoflagellates.

3.1.2 LIGHT AND PHOTOSYNTHESIS

Both the quantity (or intensity) and quality of light may effect photosynthesis. Quantitatively, light is measured in terms of energy units (e.g. calories) or units of illumination (e.g. lux). Other energy units used to measure photosynthesis are the erg and the watt–sec; the three energy units can be converted as follows:

$$1 \text{ g cal} = 4·185 \times 10^7 \text{ ergs}^* = 4·185 \text{ watt–sec.}$$

The energy flux with dimensions of energy per unit area per unit time, is much more suitable for the evaluation of photosynthesis in the sea (e.g. as g cal/cm²/min or g cal/cm²/day etc.). The g cal/cm² (i.e. without the time dimension) is called a 'langley' (1 g cal/cm² = 1 ly). Light intensity measured in units of illumination is reported as foot-candles (ft–c) or lux (1 lux = 0·0929 ft–c). The amount of radiant energy in one illumination unit depends on the spectral characteristics of the light source. Only a rough conversion from an illumination

---

* 10^7 ergs is also called 1 joule.

unit to an energy unit (or vice-versa) can be made; thus 1 lux = approx 6×10^{-6} ly/min for sunlight (Strickland, 1958), approx. 86×10^{-6} ly/min for a tungsten lamp (Hill and Whittingham, 1955); approx. 5×10^{-6} ly/min for white fluorescent lamps (Westlake, 1965).

Solar radiation available for photosynthesis in the sea changes significantly with time and depth. The subject has been reviewed by a number of authors including Strickland (1958) and Jerlov (1968). Outside the earth's atmosphere the intensity of solar radiation lies between 1·90 and 1·94 ly/min; on penetrating the atmosphere, energy is lost due to scattering and absorption, such as from water, carbon dioxide, ozone, and dust. These substances are not always present in the same amount in the atmosphere (e.g. changes in cloud cover) and consequently the mean daily values reaching the ocean surface (or land surface at sea level) in various parts of the world may vary both with seasonal changes in sun angle and atmospheric conditions. Theoretical solar radiation values, depending on latitude, season, and a clear sky have been determined (e.g. Haltiner and Martin, 1957, Fig. 7–3); actual solar radiation data are collected by a number of countries using instruments known as pyranometers (formerly, pyrheliometers). In Canada radiation is measured at 34 different stations and published monthly by the Department of the Environment. A comparison of these values and theoretical values shows, for example, that at 50°N in June and December the maximum theoretical radiation is 769 and 131 ly/day, respectively, while actual measurements at Nanaimo, B.C. (ca 50°N) during 1969 were maximum in July (622 ly/day) and minimum in December (44 ly/day).

Some solar energy is lost by true reflection, and by back scattering from particles (including foam) at the sea-surface. The actual value for surface losses varies considerably with conditions of the sea surface and sun angle; tables given by von Arx (1962) should be consulted for detailed values. On a fine day in summer, with a sun angle to the horizon of over 30°, the surface loss would be only a few percent under conditions of complete calm; this value increases to 5–17% with light winds and to over 30% for moderate to strong winds. As the sun angle decreases to less than 10°, reflection increases rapidly to over 30%. For field work, a mean value of 15% for total surface losses may be used as an approximation for conditions under which it is usually possible to carry out photosynthetic measurements.

Light penetrating into the water is absorbed and scattered selectively. The absorption of light can be expressed in terms of the extinction coefficient (k):

$$I_d = I_0 e^{-kd} \tag{22}$$

where I_d and I_0 are the radiant energy at a depth (d) and the surface, respectively. The value of (k) varies with the wavelength of light, being large for ultra-violet and infrared radiation which disappear rapidly with depth. Red light (>600 nm) is also attenuated quickly so that below a few meters, only blue–green light remains having an energy distribution between 400 and 600 nm (maximum around 480 nm). However, under turbid conditions due to particulate material in the water, blue light is selectively scattered and the spectral peak of transmitted light is moved up towards the red (maximum at ca 550 nm). A discussion of these changes in ocean and coastal waters is given by Jerlov (1968).

Light intensity strongly affects the rate of photosynthesis (usually expressed as mg C/mg Chl a/hr). Methods for the measurement of photosynthetic rate usually involve a measurement of either the carbon dioxide taken up or the oxygen produced per unit time. The ^{14}C-method first proposed by Steemann Nielsen (1952) is usually used for the measurement of CO_2 taken up since it is possible to detect very low photosynthetic rates with this method. This is particularly important in oceanic areas where photosynthetic rates are especially low and where an experimenter may only have a few hours in which to make measurements. Unfortunately some doubt may exist as to the interpretation of measurements made by this method. Because of the participation of intracellular carbonate in photosynthesis and losses from the exudation of ^{14}C-labelled organic matter (see Section 3.2.1), results by the ^{14}C-method do not necessarily measure gross, or even net, photosynthesis. It is generally assumed, however, that in open waters, the ^{14}C-method measures the rate of increase in particulate carbon (e.g. Antia *et al.*, 1963). If a seperate measurement is made of the amount of carbon lost through exudation, then the sum of the increase in particulate carbon, plus the loss of dissolved organic carbon, would be a measure of net photosynthesis. If the loss of dissolved organic carbon is small, then the ^{14}C-method will approximate net photosynthesis. The true result may be higher or lower according to the species of phytoplankton and environmental conditions. Production estimates based on oxygen evolution can be made using either an oxygen electrode or the Winkler titration technique, but both of these methods are usually an order of magnitude less sensitive than the ^{14}C-technique. Thus the oxygen technique is not particularly suitable for use in oceanic waters, but it may be used in some coastal areas, or in high latitude oceanic waters having a high density of algae. The merits and actual procedures for both methods are described in detail in the following two manuals: Strickland and Parsons (1968) and Vollenwieder (1969).

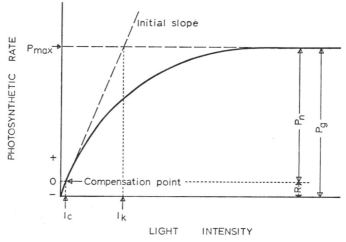

FIG. 25. Diagrammatic photosynthesis–light relationship. (P_{max}, photosynthetic maximum; I_c, light intensity at the compensation point; R, respiration; P_n, net photosynthesis; P_g, gross photosynthesis; I_k, see text).

The photosynthesis/light curve (or P vs I curve) shown in Fig. 25 is a convenient reflection of environmental effects on photosynthesis and one which can be used to diagnose certain properties of algal species, or natural samples of phytoplankton. Rabinowitch (1951), Steemann Nielsen and Jørgensen (1968) and Yentsch and Lee (1966) are recommended as references on the general interpretations of P vs I curves. From Fig. 25 it is apparent that photosynthesis increases with increasing light intensity up to some asymptotic value, P_{max} where the system becomes light saturated. The two most important properties of the curve are the slope ($\Delta P/\Delta I$) and P_{max}, which is sometimes also called the 'assimilation number' (or index). The initial slope is a function of the light reaction (see Section 3.1.1) and is not usually affected by other factors. Using the same light source, it is therefore possible to compare the initial slope of different samples to obtain an evaluation of their photosynthetic activity; a steep slope can be interpreted as high photosynthetic activity and a gentle slope represents low photosynthetic activity. P_{max}, on the other hand, is a function of the dark reaction (see Section 3.1.1) provided no environmental factors are causing photosynthetic inhibition. If other environmental factors are operative, such as low temperatures or nutrient limitation, P_{max} becomes a function of the environmental inhibitor. As a combination of the initial slope and P_{max}, Talling (1957b) proposed I_k which is the light intensity at the intersection of an extension from the initial slope and P_{max} (Fig. 25).

Figure 25 also shows the difference between total photosynthesis (or gross photosynthesis, P_g) and net photosynthesis (P_n), which is the fraction of P_g minus respiration (R). An approximate P_n can be estimated directly by the oxygen method because oxygen is consumed by respiration at the same time as it is produced through photosynthesis in the light. However, care should be taken in interpreting the oxygen method because photorespiration (see Section 3.1.1) sometimes causes an over estimation of respiration in the light. When P_g equals R, P_n is zero and the photosynthetic system is at the 'compensation point'. The light intensity at the compensation point is called the 'compensation light intensity,' I_c, and photosynthetic micro-organisms held at the compensation light intensity should theoretically show no growth. In nature, phytoplankton are subjected to continually varied light conditions and, except during the summer in extreme latitudes, to no light at all during the night. Thus at noon the amount of light reaching a cell may be above the compensation light intensity; however, over a period of 24 hours there may still be a mean light level below which the algae will decrease in weight. The compensation point is therefore best expressed on a 24-hour basis and is usually determined as the 24 hour mean light intensity in ly/day or ly/min (Strickland, 1958). From studies on natural phytoplankton populations, not excessively contaminated with other microflora or fauna, an average 24-hour compensation light intensity appears to be in the range 0·002 to 0·009 ly/min or some 3 to 13 ly/day in temperate seas (e.g. Strickland, 1958); McAllister et al., 1964; Hobson, 1966. The value is dependent on the rate of respiration which has been found to be approximately 10% of P_{max} (see Section 3.1.1). However, the basic assumption that respiration is constant in the light and dark is difficult to check and may not be valid for cells which spend a long time below the euphotic zone. For example, adaptation to photosynthesis at very low light intensities (ca 0·00014 ly/min) has been reported for phytoplankton growing under ice (Wright, 1964). Since respiration is temperature dependent, the value for I_c may be expected to be larger at higher temperatures. In field work the compensation depth can be approximated from the depth of 1% of the surface radiation; this value can either be determined using a bathyphotometer or may be approximated as twice the depth of the Secchi disk visibility (Strickland, 1958; see also Section 3.1.4). Compensation depths may vary from a few meters in turbid coastal waters to over 150 m in some tropical seas.

If micro-organisms are exposed to a strong light above the point at which they are light saturated, the P vs I curve may show a depression in photosynthetic rate. This phenomenon is named '(high) light inhibition' or 'photo-inhibition'. Photo-inhibition is not generally observed over short periods of time (e.g. 10 min.) but may result from longer exposures and may also increase in magnitude with time (Takahashi et al., 1971).

It appears from early work (e.g. Ryther, 1956) that there may exist a broad division between taxanomic groups with respect to their P vs I curves. Ryther (1956) reported P vs I curves for 15 species of marine algae representing three taxonomic groups (green algae, diatoms and dinoflagellates). Measurements were made under sunlight on a clear day at 20°C and the results showed a remarkable similarity in the photosynthetic behaviour of

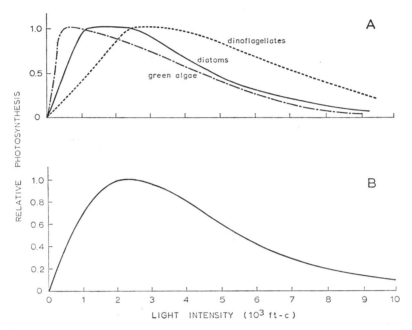

FIG. 26A. Relative photosynthesis–light curves in some marine phytoplankton. Green algae: *Dunaliella euchlora, Chlamydomonas sp., Platymonas sp., Carteria sp., Mischococcus sp., Sticho-coccus sp.,* and *Nannochloris sp.* Diatoms: *Skeletonema costatum, Nitzschia closterium, Navicula sp.* and *Coscinodiscus excentricus.* Dinoflagellates; *Gymnodinium splendens, Gyrodinium sp., Exuviaella sp.,* and *Amphidinium klebsi.* (Redrawn from Ryther, 1956).

FIG. 26B. Mean curve from Fig. 26A.

organisms within each taxonomic group, but a rather striking difference between those of different groups. A summary of Ryther's results is shown in Fig. 26; from these results it may be seen that the saturation light intensities for green algae are between 500 and 750 ft–c (or 3·3 to 4·9×10^{-2} ly/min), between 2500 and 3000 ft–c (or 16 to 20×10^{-2} ly/min) for the dinoflagellates, and at intermediate intensities for the diatoms. Photo-inhibition is apparent in all three algal groups within about 1000 ft–c (0·066 ly/min) of saturation. At intensities of 8000 to 10, 000 ft–c (52 to 66×10^{-2} ly/min), which is comparable to full sunlight, the photosynthetic rate in green algae and diatoms is only 5 to 10 per cent of that at saturation, while the photosynthetic rate for dinoflagellates is still 20 to 30% of P_{max}. The I_k of each curve in Fig. 26 is 400 ft–c (2·6×10^{-2} ly/min) for green algae, 1000 ft–c (6·6×10^{-2} ly/min) for diatoms and 2400 ft–c (16×10^{-2} ly/min) for dinoflagellates. In the sea, high I_k values are observed during the summer and in shallow algal communities, and low I_k values are observed during the winter and in deep water communities (Steemann Nielsen and Hansen, 1959a and 1961; Ichimura *et al.*, 1962).

I_k gives a measure of the radiant energy or illumination at light saturation but it does not express the photosynthetic efficiency; consequently plants or phytoplankton communities may have the same I_k values but differ appreciably in the rate of photosynthesis at I_k. In terrestial communities plants are divided into 'sun-' and 'shade-types' (Boysen Jensen, 1932) and a similar division is employed in algal communities. Thus sun-type algal communities are those which can utilize high light intensities with high efficiencies while photosynthesis in shade-type communities is generally depressed by high light intensities. However the

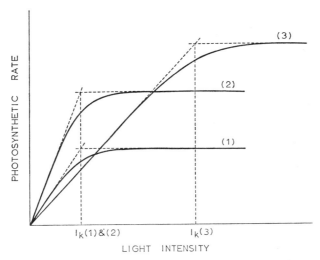

FIG. 27. Three types of P vs I curves. (1) and (2) shade type algae showing similar I_k values but with higher photosynthetic efficiency in (2) than (1). Sun-type community (3) showing lower photosynthetic efficiency than (1) or (2) at lower light intensity.

absolute photosynthetic rates of shade-type communities are usually higher than those of the sun-type communities at low light intensities. These differences in I_k values and photosynthetic efficiencies are illustrated in Fig. 27.

Table 16 shows some accumulated values for the initial slopes of P vs I curves measured under incandescent and natural light. Algal cultures of different species show values between 0·1 and 0·65 (average 0·38) mg C/mg Chl a/hr per K.lux. Those of natural samples range between 0·05 and 0·8 (average 0·4) mg C/mg Chl a/hr per K.lux. The table shows that certain species or populations are adapted to low light intensities and others to high light intensities. Differences which are sometimes found between species, or within the same community over a period of time, may be due to inactivation of cells or changes in photosynthetic mechanisms (e.g. Ichimura et al., 1968).

There are many reviews on the evaluation of P_{max} (Steemann Nielsen and Hansen, 1959a; Ichimura and Aruga, 1964; Yentsch and Lee, 1966). P_{max} is primarily influenced by environmental conditions (see Section 3.1.3) but under conditions of optimum temperature and sufficient nutrients P_{max} can be employed as the 'potential P_{max}' or 'potential photosynthetic rate'. The P_{max} of phytoplankton populations is usually measured at in situ temperature and nutrient concentrations; the results of some of these measurements, together with values for cultures, are shown in Table 17. The values range from between 1·1 to 6·2 mg C/mg Chl a/hr for cultures and from 0·1 to 6·0 mg C/mg Chl a/hr in natural samples. The data show some geographical differences in P_{max}; thus the values for polar waters are generally low, temperate waters are quite variable, and tropical waters have the highest values. Some exceptions to the general range of values have been reported; for example, Skeletonema costatum has sometimes been reported to give values during a bloom of 9·0 to 16·9 mg C/mg Chl a/hr (Hogetsu et al., 1959).

From the photosynthetic action spectra shown in Fig. 21, it is to be expected that light energy available for algal photosynthesis is restricted to wavelengths between 400 and 700 nm. Within these wavelengths the light absorbed by the phytoplankton pigments can be divided into two parts (i) the light of > 600 nm which is mainly absorbed by chlorophyll

TABLE 16. INITIAL SLOPES OF *P* vs *I* CURVES OF ALGAL CULTURES AND NATURAL POPULATIONS

Algal cultures

Species	Exp. temp. (°C)	Initial slope (mg C/mg Chl *a*/hr per K.lux)	Author
Chlorella vulgaris	20	0·50*	Steemann Nielsen, 1961
C. pyrenoidosa		0·48	Steeman Nielsen & Jørgensen, 1968
C. ellipsoidea	5–30	0·50**	Aruga, 1965b
Scenedesmus sp.	10–40	0·65**	Aruga, 1965b
Skeletonema costatum		0·52	Steemann Nielsen & Jørgensen, 1968
Skeletonema costatum	20	0·44**	Nakanishi & Monji, 1965
Chaetoceros sp.	20	0·13**	Nakanishi & Monji, 1965
Coccolithus huxleyi		0·13*	Jeffrey & Allen, 1964
Hymenomonas sp.		0·10*	Jeffrey & Allen, 1964

Natural Populations

Situation	Dominant Species	Exp. temp. (°C)	Initial slope (mg C/mg Chl *a*/hr per K.lux)	Author
Tokyo Bay	*Skeletonema costatum*	20	0·56**	Aruga, 1965b
Pond	*Synedra* sp.	10–30	0·65**	Aruga, 1965b
Pond	*Anabaena cylindrica*	10–30	0·25**	Aruga, 1965b
Lake	*Cryptomonas* sp.		*ca* 0·8	Ichimura *et al.*, 1968
Arctic		4·5–6	0·36	Steemann Nielsen & Hansen, 1959a
Tropical, temperate (summer, winter), Northern & Arctic. (general)			0·35	Steemann Nielsen & Hansen, 1959a
Kuroshio (general)			0·05	Ichimura *et al.*, 1962
Oyashio (general)			0·35	Ichimura *et al.*, 1962
Mixed regions of Kuroshio & Oyashio (general)			0·17	Ichimura *et al.*, 1962
Antarctic			0·42	Burkholder & Mandelli, 1965

*mg C/mg Chl (*a*+*b*)/hr per K.lux
**mg C/mg Chl/hr per K.lux

TABLE 17. P_{max} OF ALGAL CULTURES AND NATURAL POPULATIONS.

Algal cultures

Species	Culture conditions		Exp. temp. (°C)	light int. (K.lux)	P_{max} (mg C/mg Chl a/hr)	Author
	temp. (°C)	light int. (K.lux)				
Chlorella ellipsoidea	20		30	20	5·5**	Aruga, 1965b
C. vulgaris		3		6	1·1*	Steemann Nielsen, 1961
		30		30	3·8	
Scenedesmus sp.	20		30	10	5·0**	Aruga, 1965b
Skeletonema costatum	20	8	20	20	6·2**	Nakanishi & Monji, 1965
	20		20	13	3·6**	Aruga, 1965a
Synedra sp.	20		20	15	2·0**	Aruga, 1965b
Cyclotella		3			2·1–3·4	Jørgensen, 1964
meneghiniana		30			3·6–4·4	Jørgensen, 1964
Chaetoceros sp.	20	8	20	20	1·5**	Nakanishi & Monji, 1965
Anabaena cylindrica	20		30	10	1·7**	Aruga, 1965b
Coccolithus huxleyi	14	38	>38		>2·2*	Jeffrey & Allen, 1964
Hymenomonas sp.	14	38	>38		>1·9*	Jeffrey & Allen, 1964

Natural populations

Situation	Depth (M)	Exp. temp. (°C)	P_{max} (mg C/mg Chl a/hr)	Author
Arctic (summer)	0		1·0–1·5	Steemann Nielsen & Hansen, 1959a
Antarctic (phytoplankton)			2·3	Burkholder & Mandelli, 1965
Antarctic (ice flora)		−1·6	2·6	Burkholder & Mandelli, 1965
Antarctic (ice flora)		−1·5	0·4	Bunt, 1964b
Northern (general) no vertical stability			2·9–3·4**	Steemann Nielsen & Hansen, 1959a
Western north Pacific (summer)	10	8–20	1·4–2·0	Takahashi et al, 1972
Temperate (summer)	0		4·0–4·2**	Steemann Nielsen & Hansen, 1959a
Temperate (winter)	0		1·5**	Steemann Nielsen & Hansen, 1959a
Oyashio (general, summer)			3–6**	Ichimura & Aruga, 1964
Kuroshio (general, summer)			0·3–0·7**	Ichimura & Aruga, 1964
Bays and coastal waters near Japan (general)			2–6**	Ichimura & Aruga, 1964
Tokyo Bay (*Skeletonema* bloom, summer)		20–25	9·0–16·9**	Hogetsu et al, 1959
Tropical (general)			8·0**	Steemann Nielsen & Hansen, 1959a
Tropical Pacific (autumn)		23–27	1·1–5·2	Takahashi et al, 1972
Tropical Pacific (spring, general) nitrogen-poor water			3·15	Thomas, 1970a
Tropical Pacific (spring, general) nitrogen-rich water			4·95	Thomas, 1970a

TABLE 17. *(cont.)*

Situation	Depth (M)	Exp. temp. (°C)	P_{max} (mg C/mg Chl a/hr)	Author
Tropical Atlantic	10	20	3·0–4·0**	Yentsch & Lee, 1966
Lake (general)				Ichimura & Aruga, 1964
eutrophic			2–6**	
mesotrophic			1–2**	
oligotrophic			0·1–1**	

*mg C/mg Chl $(a+b)$/hr
**mg C/mg Chl/hr

a and (ii) the light of < 600 nm which is mainly absorbed by accessory pigments. Except in some blue–green and red algae (e.g. curve 5 in Fig. 21), the action spectra at > 600 nm of all algae are similar, while those at < 600 nm are quite different and dependent mainly on the light absorbed by accessory pigments. Since most of the light penetrating to depth is in the region 400 to 600 nm, it is quite apparent that the accessory pigments are most important as light absorbers in the ocean environment.

On cloudy days the ratio of visible light to total radiation increases (Vollenweider, 1969). Since measurements of total radiation do not differentiate between visible light and other sources of radiation, it is necessary to accommodate changes in spectral energy for evaluation of photosynthetic effect. The term 'photosynthetically active radiation' (PAR) is used to describe the portion of the spectrum that can be used for photosynthesis; on a cloudless day this can be assumed to be close to 50% of the total radiation (Strickland, 1958). If on a partly cloudy day the surface solar radiation measured with a pyranometer is greater than 50% of the total theoretical radiation, then the 50% of the total theoretical radiation is considered to be PAR; if on the other hand the radiation as measured with the pyranometer on a cloudy day is less than 50% of the theoretical value, then the total measured energy is used as PAR (Talling, 1957a; Szeicz, 1966). Alternatively if a pyranometer is equipped with a filter which allows only 400 to 700 nm radiations to pass, it can be used directly to measure PAR under all weather conditions.

Another effect of light on photosynthesis is a diurnal fluctuation in photosynthetic rate which is characterized by a high photosynthetic rate before noon and a general decline in rate during the rest of the day. This phenomenon is believed to be partly due to the destruction of chlorophyll under high light intensities and it has been observed to occur under a variety of natural conditions (e.g. Doty and Oguri, 1957; Yentsch and Ryther, 1957; Ichimura, 1960). The diurnal range of variation in P_{max} and chlorophyll content of phytoplankton (see also Section 3.1.4) has been observed to fluctuate from 2 to 8 fold in surface populations, but to decrease with depth. Also on cloudy days and during the winter, diurnal fluctuations are slight (Saijo and Ichimura, 1962). Diurnal fluctuations differ geographically; near the equator the diurnal photosynthetic range has been found to be 5 to 6 and sometimes as high as 10 to 12 fold, while at high latitudes the range is only 1 to 2 fold (Doty, 1959). Nutrient conditions have been suggested as one cause of diurnal fluctuations (Lorenzen, 1963). Vollenweider (1965) found a decrease in photosynthesis in the afternoon in lakes due, at least in part, to nutrient consumption during intense assimilation (see also Section 3.1.4).

P_{max} may change with the physiological state of algae themselves (e.g. with age) even under constant environmental conditions (Jørgensen, 1966). From experiments with synchronous cultures of *Skeletonema costatum*, it was shown that P_{max} increased in young cells (just after cell division) and reached a maximum value in full grown cells, just prior to cell division. Cell division started six hours after illumination and continued four hours after cells were placed in the dark.

From these observations it is apparent that adaptive changes in photosynthesis occur in algal cells in response to surrounding light conditions. This phenomenon is known as 'light adaptation' i.e. a physiological adjustment to surrounding conditions which has been observed to involve at least one of the following morphological or biochemical changes within the cell: (1) change in total photosynthetic pigment content, (2) change in the ratio of photosynthetic pigments, (3) change in the morphology of the chloroplast, (4) change in the arrangement of the chloroplasts, and (5) change in the availability of enzymes for the dark reaction. Specific examples of these changes have been demonstrated, for example, Fujita (1970) showed that in blue–green algae, red light induced phycocyanin synthesis and blue light increased the amount of phycoerythrin. In diatoms, the shrinking of chloroplasts and their aggregation under the influence of strong light has been observed as a reaction which is reversed under weak light (Brown and Richardson, 1968).

The time dependence of light adaptation of algae in the ocean is important in determining the day to day effects of light variation. According to Steemann Nielsen *et al.* (1962) *Chlorella* required about 40 hours to adapt to a change in light intensity from 3 to 30 K.lux at 21 °C. Algal populations taken from the surface of strong vertically mixed water masses off Friday Harbor, Washington, U.S.A., took three days to adapt to low light conditions of about 5% of the surface illumination (Steemann Nielsen and Park, 1964). If algae are kept in the dark, there is a gradual loss in photosynthetic activity which in the case of *Nannochloris* amounted to 50% of the photosynthetic activity in about 40 hours at 20 °C (Yentsch and Lee, 1966).

3.1.3 NUTRIENT AND TEMPERATURE EFFECTS ON PHOTOSYNTHESIS AND THE GROWTH REQUIREMENTS OF PHYTOPLANKTON

Photosynthesis of algae is also controlled by factors other than light, such as nutrients and temperature, and to a lesser extent a variety of factors, such as pH and salinity. Liebig (1840) postulated a simple rule for the effect of various factors on yield (i.e. net photosynthesis, P_n). His statement was that "growth of a plant is dependent on the minimum amount of foodstuff presented" and this has come to be known as "Liebig's law of the minimum". Sixty years after Liebig's work, Blackman (1905) suggested that a generalized form of this law could be applied to photosynthesis in order to explain field and experimental observations. He took from Liebig the idea that the rate of a biological process (in this case, photosynthesis) is determined, under given conditions, by a single limiting factor. However, in addition to the supply of material ingredients (the only kind of factors with which Liebig was concerned), Blackman considered also light intensity and temperature as limiting factors. He suggested that the rate of photosynthesis increased with an increase in the value of any one of these factors (F_1), as long as the particular factor was rate limiting, and that it ceased to be dependent on F_1 when one of the other factors ($F_2, F_3 \ldots$) became limiting. In other words, the plot of photosynthesis, P_n, versus a variable F_1 (at constant values of all other kinetic factors) was postulated by Blackman to have the shape 1, 2, 3 in Fig. 28. In actual fact the curve approaches the maximum asymptotically, without a sudden break as indicated by the point 2. It should be noted also that an excessive amount of the factor, F_1, eventually causes a depression in P_n. Assuming F_1 to be light intensity, the P_n vs F_1 relation is the same as the P vs I curve shown before in Section 3.1.2. In order that other factors, F_2, F_3 ... should conform to the same rule, it is apparent that for different values of a second parameter (F_2), photosynthesis (P_n) as a function of F_1 will be represented by a sequence of solid lines, such as 1.2.3, 1.4.5, 1.6.7, in Fig. 28. These coincide at low F_1 values (part 1.2 in the figure) but are distinguished at higher F_1 values by the position at which the ascending part of the

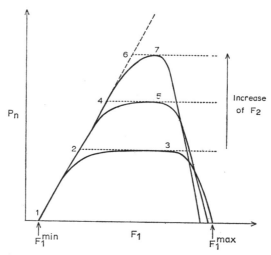

FIG. 28. Changes in photosynthetic rates, P_n, with a change in two environmental factors, F_1 and F_2, varied independently.

curve becomes horizontal. This value is determined by the second factor (F_2), which causes an increase in P_n as represented by levels 3, 5 and 7. The horizontal plateau in Fig. 28 does not extend indefinitely, however, and P_n declines when F_1 causes an inhibition. The initial slope, which is dependent on F_1, tends to be similar for the sequence of curves but the decline may be more rapid with an increase in F_2. The maximum and minimum values F^{max} and F^{min}, in Fig. 28 are called the 'upper lethal limit' and the 'lower lethal limit' respectively, and these can be generally recognized in applying any factor to a growth process. This gives a numerical evaluation to descriptive terms which have been used to express a relative degree of tolerance. Thus the series of expressions which have come into general use in ecology (see Section 1.1) and which utilize the prefixes 'steno-' meaning narrow and 'eury-' meaning wide, with reference to the tolerance of a process, or an organism, can be given definite meaning in terms of Fig. 28. Furthermore, the ascending and descending slopes, together with the width and height of the plateau, are also important physiological characters to consider in the ecological adaptation of certain algal populations to a given aquatic environment.

An example of the relationship between the seasonal photosynthesis of mixed algal populations, and different phosphate concentrations and temperatures, is given by Ichimura (1967) for phytoplankton in Tokyo Bay (Fig. 29A). From these results it is apparent that phosphate regulation on P_{max} is governed by temperature; the relationship is quite similar to the Blackman-type limitation shown in Fig. 28, although the higher concentrations of phosphate were not sufficient to cause a break in the curve. Fig. 29B from Aruga (1965b) serves as a second example of the general relationship in Fig. 28. In this case the influence of

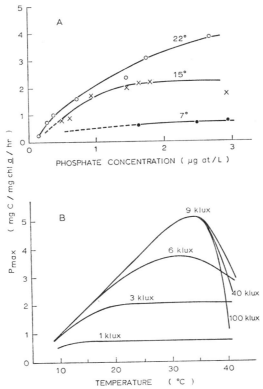

FIG. 29. Photosynthesis regulation by phosphate (A) and temperature (B). (A) natural populations taken from Tokyo Bay, (redrawn from Ichimura, 1967); (B) cultured *Scenedesmus sp.*, (redrawn from Aruga, 1965a).

light and temperature on photosynthesis gives a more complete Blackman-type response including depression of P_{max} at high light intensities. While these results were obtained for local populations which had been reconditioned to a specific environment, it is also known that populations can adapt to some extent to different environments. Steemann Nielsen and Jørgensen (1968) showed, for example, that a phytoplankton population could become adapted to a new temperature regime within a few days. For some species, however, temperature adaptation has to be made in a series of small steps (e.g. 5 °C changes) in order to exclude harmful effects caused by a sudden change in temperature. Experimental data on the effect of temperature on photosynthesis are generally scarce but some studies have been made in the Antarctic (Bunt, 1964a, b) and in the Pacific (Aruga *et al.*, 1968; Ichimura *et al.*, 1962); the latter results indicated that temperate Pacific phytoplankton had their highest P_{max} at about 20°C in spite of *in situ* temperatures which varied between -0.9 and 17.9 °C.

P_{max} measured at *in situ* temperatures is a good indicator of what kinds of factors (other than light) are limiting photosynthesis. As an example, seasonal variations in P_{max} from Tokyo Bay are shown in Fig. 30. From the results it is apparent that from October to June, temperature regulates photosynthesis; during July to September P_{max} is depressed by the lack of nutrients which increase again due to autumn mixing in October. On the other hand, the potential photosynthetic rate, which is measured under optimum temperature and nutrient conditions, is constant throughout the year except in February and August. Low

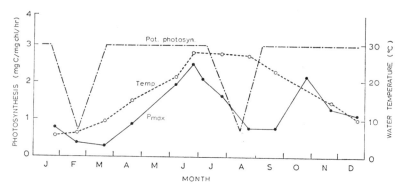

FIG. 30. Seasonal variations in temperature, P_{max} and potential photosynthetic rate in Tokyo Bay (redrawn from Ichimura and Aruga, 1964).

potential photosynthetic rates in these months have been attributed to a changing phytoplankton flora during the early spring and late summer. In high latitudes where there are low temperatures and high nutrient concentrations, P_{max} will be largely regulated by temperature. Phytoplankton communities near river mouths may also be regulated by temperature throughout most of the year (Ichimura, 1967). On the other hand, P_{max} in tropical and sub-tropical communities is more likely to be limited by nutrients.

Eppley (1972) has summarized the effect of temperature on algal growth rate. By plotting growth rate data, obtained by different people, *vs* temperature, he found an empirical relation for the maximum growth rate of algae over the temperature range between 0 and 40 °C, under conditions of continuous illumination:

$$\log_{10} \mu = 0.0275 \, t - 0.070 \qquad (23)$$

where μ is the maximum possible growth rate in doublings per day; (see Section 3.1.4), and t is temperature in degrees Celsius. The equation was deduced from data on algal cultures which included a wide variety of taxonomic groups, cells with different complements of photosynthetic pigments, and diverse morphologies. The growth rate of each algal species (or clone) can be fitted to the equation over a limited temperature range; growth ceases at temperatures above a supraoptimal point, which is a characteristic of the species or clone (Eppley, 1972).

Algae require certain elements for their growth. Some of these elements, such as C, H, O, N, Si, P, Mg, K, and Ca are needed in relatively large amounts and are often known as 'macro-nutrients'. Other elements are required in very small amounts and are referred to as 'micro-nutrients' or 'trace elements'; Hewitt (1957) has listed 10 trace elements required by green plants as follows: Fe, Mn, Cu, Zn, B, Na, Mo, Cl, V, and Co. Most of these elements are contained in sufficient abundance for algal growth in sea water; however, nitrogen and phosphorus may often limit plant growth during the summer or in tropical and sub-tropical latitudes throughout the year.

As an inorganic carbon source, algae can use free carbon dioxide, bicarbonate, and carbonate. These three are measured as 'total carbon dioxide' and the amount is expected to be *ca* 90 mg CO_2/l in offshore pelagic waters. At this concentration phytoplankton photosynthesis is not limited by the total amount of carbon dioxide. For example, Talling (1960) did not find any stimulation or reduction in photosynthesis following carbonate addition to a culture of *Chaetoceros affinis* in natural sea water. When phytoplankton grows vigorously, such as in a 'bloom', the total carbon dioxide content of sea water (determined as the partial pressure) may show a negative correlation with chlorophyll concentration in temperate waters (Gordon *et al.*, 1971).

Nitrogen and phosphorus are of major importance as metabolites, and their concentration should always be considered first in determining possible limitations in primary production. The ratio in which nitrogen and phosphorus are taken up from sea water was studied by Redfield (1934) and Cooper (1937), who concluded that both in natural populations and cultures, the atomic ratio of cellular N : P was about 16 : 1 (see also Table 11). However, under some circumstances, depending on species and nutrient availability, different N:P ratios are encountered. For example, Ketchum's (1939b) N : P ratios for some representatives of the Chlorophyceae were about 7 : 1.

In the oceans nitrogen exists mainly as molecular nitrogen and as inorganic salts, such as nitrate, nitrite and ammonia. The usual range of concentration for these salts in sea water is 0·01 to 50 µg at/l for nitrate, 0·01 to 5 µg at/l for nitrite and 0·1 to 5 µg at/l for ammonia. Molecular nitrogen is fixed by certain blue–green algae [e.g. *Calothrix* sp. (Allen, 1963) and *Trichodesmium* spp. (Dugdale *et al.*, 1964; Goering *et al.*, 1966; Dugdale and Goering, 1967)], yeasts [e.g. *Rhodotorula* sp. (Allen, 1963)] and bacteria [e.g. *Azotobacter* and *Closteridium* (Pshenin, 1963)]. However, compared with freshwater microorganisms, nitrogen fixers in the marine environment have not been clearly defined. *Trichodesmium* can also actively utilize both ammonia and nitrate but since the tropical habitat of *Trichodesmium* is usually relatively poor in nitrate, nitrogen fixation, together with ammonia utilization, are probably the most important sources of algal nitrogen in such environments (Goering *et al.*, 1966). Most other algae have no ability to fix molecular nitrogen and must utilize inorganic nitrogen salts. Algae generally show a preferential utilization for nitrate, nitrite, and ammonia (e.g. Guillard, 1963); some exceptions are found among the green algae (Braarud and Føyn, 1931) and a few flagellates (e.g. Strickland, 1965). Of the three inorganic salts, ammonia is utilized in preference to nitrate (Morris and Syrett, 1963; Dugdale and Goering, 1967; Eppley *et al.*, 1969a). Ammonia is used directly for amino acid synthesis through transamination, but nitrate and nitrite must be reduced before being utilized by the cell. Reduction of nitrite appears to be carried out in the light (Hattori, 1962a; Grant, 1967; Eppley and Coatsworth, 1968) using photosynthetically reduced ferredoxin to reduce nitrite to ammonia (Hattori and Myers, 1966). Nitrate is reduced by the action of the enzyme, nitrate reductase, which is not dependent on light; the occurrence of this enzyme is induced by the presence of nitrate (Hattori, 1962a and b), but not by nitrite (Eppley *et al.*, 1969a). In some areas the principal source of nitrogen for phytoplankton may differ with depth in the same water column. For example, in the eastern subtropical Pacific Ocean off the continental shelf of Mexico, a pronounced discontinuity layer often exists at between 40 and 60 m within the euphotic zone, and phytoplankton in this layer use nitrate as their major nitrogen source; those in the overlying nitrate-impoverished water 0–40 m use ammonia which is derived largely from nitrogen recycled through zooplankton (Goering *et al.*, 1970). Under culture conditions some algae have been shown to be capable of using dissolved organic nitrogen compounds including hydroxylamine, casein, amino acids, urea, and uric acid (e.g. Loeblich, 1966; Holm-Hansen, 1968). However, little is known at present about the metabolism of these compounds in natural sea water.

Phosphorus occurs in sea water in three principal phases: dissolved in organic phosphorus, dissolved organic phosphorus, and particulate phosphorus. However, the existence of these forms is quite complex and special analytical procedures are needed to show the different forms of phosphorus in sea water (e.g. Strickland and Parsons, 1968). Phytoplankton normally satisfy their requirement for this element by direct assimilation of dissolved inorganic phosphorus (orthophosphate ion) and sometimes by utilizing dissolved organic phosphorus (Provasoli and McLaughlin, 1963; McLaughlin and Zahl, 1966). In polluted water, polyphosphate, which is inorganic, and organic soluble phosphorus detected as orthophosphate after acid hydrolysis, may be present in appreciable amounts. Some coastal algae, such as *Skeletonema costatum* and *Amphidinium carteri*, can use polyphosphate as a phosphorus source in the presence of excess nitrate. These species appear to be able to hydrolyse an external supply of polyphosphate more rapidly than it is utilized by the cells with the result that orthophosphate may accumulate in the surrounding water (Solórzano and Strickland, 1968). Phosphorus absorbed into a cell becomes part of the structural component of a cell (e.g. in poly-P-RNA) and is in part continually turned over in the energetic processes of organisms (e.g. an adenosine di- and tri-phosphate). In this sense the role of phosphorus as a metabolite is quite different from nitrogen because nitrogen is used primarily as a structural component of cells and not directly in the energy cycle of the cell.

Diatoms and silicoflagellates take up a great amount of dissolved silicon and deposit it as hydrated silica to form their elaborately patterned valves. The concentrations of dissolved silicon are generally high in coastal and deep pelagic waters and low in surface waters away from the influence of estuaries. Diatoms grown in a medium low in silicate become silicon deficient but may remain viable for several weeks in the dark. However, on exposure to bright light they photosynthesize for a limited period and soon die (Riley and Chester, 1971). Diatom blooms deplete the surface waters of silicate and Menzel *et al.* (1963) have suggested that this is the chief mechanism leading to a species succession of non-silicious flagellates following the spring diatom bloom in sub-tropical waters.

Ferric iron as a hemin complex known as cytochrome plays an important role in cellular respiratory processes; in addition, the iron complex called ferredoxin is an essential part of the light reaction in photosynthesis (Section 3.1.1). Natural sea water contains from 1 to 60 μg/l of iron but only some part of the iron detected analytically seems to be available for algal metabolism. A shortage of iron for phytoplankton growth has been demonstrated in the pelagic waters of the Sargasso Sea (Menzel and Ryther, 1961; Menzel *et al.*, 1963) and off the coast of Australia (Tranter and Newell, 1963). High concentrations of iron in sea water are generally associated with river run off (see e.g. Williams and Chan, 1966). Ryther and Kramer (1961) compared the iron requirements of a number of coastal and oceanic species of phytoplankton and found that coastal species could not grow at the low iron concentrations at which oceanic species would grow. In a reversal of this situation Strickland (1965) has suggested that the growth of some oceanic species of phytoplankton in coastal waters may be inhibited by relatively high trace element concentrations. In some instances differences in the micro-nutrient concentration may be governed more by their availability through chelating agents than through their absolute concentration. The ability of sea water samples to support the growth of phytoplankton has been observed by a number of authors to be quite variable (e.g. Johnston, 1963; Smayda, 1964 and 1970b; Barber *et al.*, 1971). These studies have led to a general conclusion that subtle differences in the micro-nutrient composition of sea water, including the presence of organic substances, may play an important role in determining the total productivity of the water and the diversity of species present. The chemical identity of these substances remains obscure although some work on the origin of soluble organic substances (e.g. Prakash and Rashid, 1968; Pratt, 1966) has shown that humic acids, and the exudates from other phytoplankton, may inhibit or enhance the growth of algae.

The total spectrum of inorganic nutrients needed for phytoplankton growth can be illustrated from the work of Provasoli and others (e.g. Provasoli *et al.*, 1957), who developed several types of artificial sea water media for the growth of phytoplankton; one of these media which has been used extensively in phytoplankton culture experiments is given in Table 18. The ingredients illustrate the need for trace elements and organic compounds, such as the vitamins and a chelating agent (Na_2EDTA). The vitamin requirement of some organisms should be considered as a special case of autotrophy; the name 'auxotroph' is used to

TABLE 18. SYNTHETIC SEA WATER MEDIUM FOR THE GROWTH OF PHYTOPLANKTON (ASP 2 FROM PROVASOLI *et al.*, 1957).

Compound	wt/100 ml	Compound	wt/100 ml
NaCl	1·8 g	TRIS[2]	0·1 g
$MgSO_4 \cdot 7 H_2O$	0·5 g	B_{12}	0·2 mg
		Vitamins[1]	1·0 ml
KCl	0·06 g	Na_2 EDTA[3]	3·0 mg
Ca (as Cl^-)	10 mg	Fe (as Cl^-)	0·08 mg
		Zn (as Cl^-)	15·0 mg
$NaNO_3$	5 mg	Mn (as Cl^-)	0·12 mg
		Co (as Cl^-)	0·3 mg
K_2HPO_4	0·5 mg	Cu (as Cl^-)	0·12 mg
$Na_2SiO_3 \cdot 9 H_2O$	15 mg	B (as H_3BO_3)	0·6 mg

[1]Vitamins: 1 ml contains 0·05 mg thiamine HCl, 0·01 mg nicotinic acid, 0·01 mg pantothenate, 1·0 mg *p*-aminobenzoic acid, 1·0 mg biotin, 0·5 mg inositol, 0·2 mg folic acid, 0·3 mg thymine.
[2]TRIS buffer (2-amino-2-hydroxymethyl-propane-1,3 diol)
[3]EDTA chelating agent (Ethylenediaminetetra-acetic acid disodium salt)

describe organisms which have a physiological growth requirement for one or more organic compounds but which derive their carbon from CO_2 and energy from light. The need for such growth factors should not be confused with the use of organic materials as an energy source, which is considered in the next section under 'heterotrophy'. Provasoli (1958) reviewed some of the growth factor requirements of the algae and showed that among the commonest groups of phytoplankton, the Dinophyceae exhibited the most extensive auxotrophic requirements which often included a need for thiamine, biotin, and vitamin B_{12}. Many phytoplankton do not require any of the organic growth supplements shown in Table 18. In studies on truly autotrophic organisms it should only be necessary to add the inorganic compounds shown in Table 18, together with a buffer (Tris) and the chelating agent (Na_2EDTA), neither of which are used directly by the cells. Both substances may be necessary in cultures, however, since the small volumes of culture vessels are unrepresentative of the natural seawater environment, and the pH and chelating properties in culture vessels can change much more drastically than in the sea. In addition, for large standing stocks of phytoplankton it may be necessary to bubble CO_2 enriched air through a culture medium.

3.1.4 PHOTOSYNTHESIS AND GROWTH OF PHYTOPLANKTON IN THE SEA

Under natural conditions, photosynthesis of phytoplankton is regulated spatially and temporally by several environmental factors. Since the general effect of these environmental factors on photosynthesis is known, *in situ* photosynthesis can either be measured directly, or indirectly by a mathematical combination of data on environmental changes and physiological responses of the phytoplankton.

The direct measurement of photosynthesis is used extensively. In this method, samples are collected from various depths in the euphotic zone and each sample is used to fill at least one clear glass bottle and one opaque glass bottle, replicates at each depth being filled as desired. The bottles are innoculated with radioactive bicarbonate (Steemann Nielsen, 1952) and a set of light and dark bottles are then returned to the same depth from which the sample was taken. The vertical spacing of the samples to be collected and exposed depends mainly on the depth to which the water column is illuminated. For example, the sampling depths may be chosen as follows: 100, 50, 25, 10, 5, and 1% of the surface illumination. These depths can either be determined by using a known value for the extinction coefficient [eqn. (22) in Section 3.1.2], or approximated from the Secchi disc depth using the relationship, $k_e = 2 \cdot 1/T$ (Jones and Wills, 1956) where k_e is the extinction coefficient (m^{-1}) and T is the depth of the Secchi disc in metres.

The light and dark bottles remain suspended from a buoy during the period of the incubation and the carbon dioxide taken up in the light bottle minus the same value for the dark bottle, is considered to be the amount of particulate organic carbon produced by photosynthesis. This value is probably somewhere between a measure of net and gross photosynthesis, but it is often referred to as net photosynthesis if factors, such as the exudation of soluble organic carbon, are ignored.

The suspension of bottles in the water column is conveniently carried out during half a day (i.e. from dawn to mid-day, or mid-day to dusk). The fraction of the daily radiation occurring during the period of the incubation can either be measured with a pyranometer or calculated using a formula, such as that proposed by Vollenweider (1965) or Ikushima (1967). The transformation of short incubation periods into daily rates (Platt, 1971) does not, however, take into account that photosynthesis may decline during incubation. This may be due to a diurnal photosynthetic rhythm (Vollenweider and Nauwerck, 1961), or to qualitative and quantitative changes in the enclosed population (Ichimura and Saijo, 1958). Physiological reasons for the latter have not been studied extensively, but losses due to damage during manipulations could well be a contributing factor (e.g. to sensitive flagellates). Changes in the daily photosynthetic rhythm have been studied more extensively and it appears (Lorenzen, 1963) that maximum photosynthetic rate (per unit of chlorophyll *a*) usually occurs before noon and about 70% of the total daily photosynthesis at the surface is carried out in the morning (i.e. between dawn and mid-day). However, at different depths in the water column the amount of light will be a function of sun angle; consequently, the period for which light is available for photosynthesis is partially a function of depth and this tends to reduce the diurnal rhythm in the water column as compared with observed changes in surface samples. For the whole water column, Vollenweider and Nauwerck (1961) found that photosynthesis in the morning was 54 to 62% of the daily photosynthesis per unit area.

Another disadvantage to short exposures is that it ignores the influence of temporary changes in weather conditions as well as diurnal changes in illumination. Thus in spite of the physiological difficulties of long exposures mentioned above, half or full day incubations are usually performed, and correction factors are employed for day-to-day and seasonal differences in illumination and day length.

In many cases it is inconvenient to keep bottles suspended *in situ* for half or one day periods. Such situations arise due to the limited availability of ship time at one location or in some near shore environments which are not readily accessible. In order to avoid this difficulty, extensive use has been made of various simulated *in situ* methods involving an experimental approach and a mathematical interpretation of the results. In this technique, instead of suspending bottles in the sea, each light and dark bottle is put under a suitable light filter adjusted to the light transmission at the depth at which the sample was taken. The sample is exposed to sunlight in a temperature controlled incubator (Jitts, 1963; Vollenweider, 1969; Kiefer and Strickland, 1970).

The indirect method involves measuring the photosynthetic response of phytoplankton to different environmental factors; all results are then combined and integrated mathematically (Ichimura, 1956b; Talling, 1957b), or graphically (Ryther, 1956) in order to determine the total amount of carbon dioxide fixed in the water column. Since light intensity is the most changeable parameter under field conditions, spatial and temporal changes in radiation, and the response of phytoplankton, are the most important experimental results. For this technique a water sample is usually taken from just under the sea surface (or from specific light depths if time permits) and then photosynthesis is measured at the *in situ* temperature and under a series of sunlight illuminations (e.g. 100, 50, 25, 10, 5, and 1% of the surface illumination). Exposure to sunlight is made for a few hours (e.g. 3 hr.) around local apparent noon. Other light sources, such as an incandescent lamp showing high colour temperature (e.g. 3200 K), or a day-light type fluorescent tube, may be used conveniently.

Photosynthetic response of phytoplankton to light of different intensities (e.g. P vs I curves) has already been discussed in Section 3.1.2 (Figs. 24, 25, and 26). The P vs I curve used in these experiments is then approximated using a mathematical equation; several equations have been proposed for this approximaton and the simplest is probably that of hyperbola (Tamiya *et al.*, 1953):

$$P_g = \frac{bI}{1+aI},\qquad(24)$$

where P_g is gross photosynthesis (usually expressed as mgC/mg Chl a/hr), I is the ligh intensity in suitable energy or illumination units, and a and b are constants. When $I \rightarrow \infty a$ $P_g = b/a = P_{max}$ (the maximum photosynthesis of the curve); eqn. (24) can be replaced (Tominaga and Ichimura, 1966) by:

$$P_g = P_{max} \cdot \frac{aI}{1+aI} \quad \text{or} \quad P_g = \frac{P_{max}I}{(1/a)+I}.\qquad(25)$$

The first function is similar to an equation known as Smith's equation (Smith, 1936) in which $\sqrt{1+(aI)^2}$ is used instead of $(1+aI)$. The second function is in a form similar to the Michaelis–Menten expression for enzyme reactions (see Section 3.1.5).

Under high light intensities photosynthesis is inhibited and this is not apparent in eqn. (25). Several authors have proposed a solution to this difficulty (Steele, 1962; Vollenweider, 1965. In the solution proposed by Steele (1962), photo-inhibition was expressed in the form:

$$P_g = \alpha \cdot I \cdot P_{max} \cdot e^{(1-\alpha I)}\qquad(26)$$

where α is a constant. Unfortunately, it is difficult to fit this equation to experimental data and in order to solve this problem two different constants ($m \neq n$) are introduced in eqn.

(26) (Vollenweider, 1965):

$$P_g = m \cdot I \cdot P_{max} \cdot e^{(1-nI)}.$$ (27)

For the purpose of a better fit, Vollenweider (1965) proposed an equation which is a modification of eqn. (25).

$$P_g = P_{max} \cdot \frac{aI}{\sqrt{1+(aI)^2}} \cdot \frac{1}{\{\sqrt{1+(\beta I)^2}\}^n}$$ (28)

where a, β and n are constants. This equation can be used extensively to fit observed P vs I curves. When $n = 1$ and $\beta = 0$ the curve shows the same form as eqn. (25); when $\beta > 0$ the curve shows the same form as eqn. (27). However, the weakness in the practical application of eqn. (28) is in difficulties resulting from its integration with time or area. There is no general solution to such an integral except in some specific cases; the use of a computer makes possible an approximation of the integration (Fee, 1971). One other solution to the approximation of P vs I curves is a polynomial curve using multiple regression analysis of data.

The constants in all of the above equations have eco-physiological meaning; the b and b/a in eqn. (24) reflect the initial slope and P_{max} of a P vs I curve, and represent the physiological response of algal populations, or the ecological adaptation of populations to light intensity. Furthermore, the α in eqn. (26) is a reflection of photo-inhibition; the n and β of eqn. (28) are also related to photo-inhibition and the effects of changes in these constants on curve approximation have been shown graphically by Vollenweider (1965) and Fee (1969). Knowledge of how these constants change with environmental conditions and algal species based on many P vs I curves taken from different areas and at seasons, may eventually provide suitable P vs I curves without actual measurements of photosynthesis.

As has already been mentioned in this section, photosynthetic rate changes diurnally (see also Section 3.1.2). For example, evening rates at optimum light intensity may drop as low as one tenth of that measured during the early morning hours (e.g. Doty, 1959). In some waters this decrease may be a function of diurnal nutrient depletion (e.g. in the tropics), or of saturation of photosynthetic dark reactions.

Apart from the physiological depression in photosynthetic response, the overall diurnal response of photosynthesis is dependent on changes in the light intensity. In general, the diurnal change of incident light intensity on a horizontal plane can be given by empirical equations (Vollenweider, 1965; Ikushima, 1967). According to Ikushima, for a fine day

$$I_t = I_{max} \cdot \sin^3 (\pi/D)t$$ (29)

where I_t (expressed as a unit of photosynthetically available light energy or illumination) is the light intensity at a given time, t, on a given day and I_{max} is the light intensity when the altitude of the sun is highest at local apparent noon; D (hr) is the day length. If the relationship of light penetration in a water column (Section 3.1.2) is put into the I_{max} of eqn. (29), light intensity at a given depth and time can be calculated for a fine day as

$$I_{t \cdot d} = I_{0 \cdot max} \cdot \sin^3 (\pi/D)t \cdot e^{-kd}$$ (30)

where $I_{0 \cdot max}$ is the light intensity at the surface for the highest altitude of the sun, k is the average extinction coefficient of photosynthetically available light (m^{-1}), and d is the depth in meters. For overcast days the equations given above should be modified by taking the $\sin^2$ instead of the $\sin^3$; this effectively reduces the total radiation while keeping the same general shape of the diurnal curves.

For combining a P vs I curve (physiological response) and light conditions (environmental response), eqn. (30) is inserted into the I of eqn. (24) or (25) and an equation for calculating gross photosynthesis at a given depth (d) and time (t), $P_{g \cdot t \cdot d}$ is obtained. For example, from eqn. (25) and (30):

$$P_{g \cdot t \cdot d} = P_{g \cdot t} \cdot \frac{a \cdot I_{0 \cdot \max} \cdot \sin^3 (\pi/D)t \cdot e^{-kd}}{1 + a \cdot I_{0 \cdot \max} \cdot \sin^3 (\pi/D)t \cdot e^{-kd}} . \tag{31}$$

Instead of using empirical equations [eqn. (29) and (30)], actual measurements of solar radiation can be put directly into an equation [e.g. eqn. (25)] for a P vs I curve if they are available (Section 3.1.2), and then the following calculations can be continued.

By integrating eqn. (31) with time ($t = 0 \rightarrow D$), daily gross photosynthesis (usually in mg C/mg Chl a/day), at a given depth (d) is calculated as

$$P_{g \cdot d} = \int_0^D P_{g \cdot t \cdot d} \cdot dt. \tag{32}$$

By integrating eqn. (32) with depth ($d = 0 \rightarrow \infty$), daily gross photosynthesis (usually mg C/day/m²) is calculated as

$$P_g = \int_0^\infty P_{g \cdot d} \cdot dd. \tag{33}$$

For determining the growth of phytoplankton under natural conditions, net photosynthesis should be considered instead of gross photosynthesis. Net photosynthesis can be calculated by subtracting respiration, R(mg C/mg Chl a/hr), from gross photosynthesis as follows:

Hourly net photosynthesis, $P_{n \cdot t \cdot d}$;

$$P_{n \cdot t \cdot d} = P_{g \cdot t \cdot d} - R. \tag{34}$$

Daily net photosynthesis, $P_{n \cdot d}$;

$$P_{n \cdot d} = P_{g \cdot d} - 24R \tag{35}$$

where respiration rate of phytoplankton is assumed to be constant throughout the day. The depth at which $P_{n \cdot t \cdot d} = 0(P_{g \cdot t \cdot d} = R)$ or, $P_{n \cdot d} = 0(P_{g \cdot d} = 24 R)$ is called the hourly and daily compensation depth, respectively. The hourly compensation depth will change during the day and be maximum at noon and zero during darkness; the 24 hour compensation depth will change with season. Unfortunately, it is not often clear in the literature which compensation depth is being reported.

Daily photosynthesis obtained above is the daily rate when 1 mg Chl a/m³ is distributed homogeneously in the euphotic zone; this is not the actual daily photosynthesis of the water column since the amount of chlorophyll will vary in time and space. In order to obtain the actual photosynthesis, $\mathbb{P}_{g \cdot d}$, the $P_{g \cdot d}$ must be multiplied by the amount of phytoplankton biomass, S (usually expressed as mg Chl a/m³):

$$\mathbb{P}_{g \cdot d} = P_{g \cdot d} \times S \tag{36}$$

$$\mathbb{P}_{n \cdot d} = P_{n \cdot d} \times S. \tag{37}$$

$\mathbb{P}_{n \cdot d}$ is sometimes referred to as ΔN, which represents the daily net increase in standing stock, N. Hourly photosynthesis in a P vs I curve is sometimes expressed on the basis of a unit volume of water, such as mg C/m³/hr. In such a case, however, the curve can only be applied to the depth from which the sample for the measurement of P vs I was taken, and at all depths, only in the season when the chlorophyll a is distributed homogeneously, such as during active vertical mixing. When chlorophyll a is not uniformly distributed, the expression for photosynthesis per unit of chlorophyll a must be obtained by multiplying the photosynthesis by the actual amount of chlorophyll a at different depths in the water column. Finally the depth integration of $\mathbb{P}_{g \cdot d}$ can usually be done with a planimeter, or by graphic integration.

In determining an average growth rate, μ, for the phytoplankton over short intervals, the biomass (usually expressed as mg C/m^3) is taken initially as N_0. The measured daily increase of phytoplankton, assumed to represent daily net photosynthesis (ΔN) on a unit carbon base instead of chlorophyll a, is then added to the biomass after a day's growth (t). For transforming daily net photosynthesis $P_{n \cdot d}$ (mg C/mg Chl a) into ΔN, $P_{n \cdot d}$ must first be converted to cellular carbon by multiplying by the amount of chlorophyll a in the water [eqn. (37)]. Then

$$\Delta N = N_t - N_0 \tag{38}$$

where N_t is the biomass after t days and the average growth rate is calculated as

$$\mu = 1/t \cdot \log_e [(N_0 + \Delta N)/N_0]. \tag{39}$$

The average growth rate is often expressed in terms of 'doublings of algal biomass per day'. For this purpose the base of the logarithm in eqn. (39) should be replaced by 2 instead of e (2·7183).

TABLE 19. AVERAGE GROWTH RATE OF PHYTOPLANKTON IN THE EUPHOTIC ZONE FOR VARIOUS REGIONS. TEMPERATURES INDICATED ARE FOR THE SURFACE OR THE AVERAGE IN THE MIXED LAYER. (AFTER EPPLEY, 1972).

Location	Temp. (°C)	Growth rate (doublings/day)		Reference
		measured	max. expected*	
Nutrient-poor waters				
Sargasso Sea	—	0·26	—	Riley *et al.* (1949)
Florida Strait	—	0·45	—	Riley *et al.* (1949)
Off the Carolinas	—	0·37	—	Riley *et al.* (1949)
Off Montauk Pt.	—	0·35	—	Riley *et al.* (1949)
Off Southern California				
July, 1970	20	0·25–0·4	1·5	Eppley *et al.* (1972)
Apr.–Sept., 1967	12–21	0·7 av.	0·9–1·6	Eppley *et al.* (1972)
Nutrient-rich waters				
Peru Current				
Apr., 1966	17–20	0·67 av.	1·5	Strickland *et al.* (1969)
June, 1969	18–19	0·73 av.	1·4	Beers *et al.* (1971)
Off S. W. Africa	—	1·0 av.	—	Calculated from Hobson (1971)
Western Arabian Sea	27–28	>1·0 av.	2·4	Calculated from Ryther and Menzel (1965)

*From eqn (23) (Section 3.1.3) assuming the maximum growth rate, μ, will be one-half the value calculated as expected if daylength is 12 hours and μ is directly proportional to the number of hours of light per day

Average growth rates of phytoplankton in the euphotic zone for various regions are shown in Table 19. The growth rates, measured in an experimental bottle, represent the potential rate under given environmental conditions on a particular day. The actual growth rate *in situ* may be smaller or larger than the potential rate. Smaller *in situ* growth rates may result from losses of phytoplankton cells in the water column due to sinking, drifting, dying, zooplankton feeding, vertical mixing, and the movement of phytoplankton themselves. Larger *in situ* growth rates may result in nutrient-depleted environments since the exclusion of animals from the incubation bottles may decrease nutrient regeneration.

Of these effects, the most pronounced in temperate latitudes is the depth to which the water column is mixed. Thus, under conditions of storm activity, or tidal mixing in coastal areas, it is apparent that phytoplankton cells may be mixed down below the euphotic zone and that the time that the cells spend in the aphotic zone before being mixed back to the euphotic zone, may be sufficient to result in no net production; the period of darkness below the aphotic zone having resulted in a loss through respiration of the carbon gained through photosynthesis in the euphotic zone. In this sense the euphotic zone is defined strictly as the depth at which the photosynthesis of a plant cell equals its respiration; the depth of the euphotic zone is known as the compensation depth (D_c) and the light intensity at this depth is the compensation light intensity (I_c).

The depth to which plants can be mixed and at which the total photosynthesis for the water column is equal to the total respiration (of primary producers) is known as the "critical depth". The concept of a critical depth was first suggested by Grann and Braarud (1935) and developed into a mathematical model by Sverdrup (1953). The following description of the model is taken from Sverdrup's original paper, with a few modifications. In particular, Sverdrup reduced the incident solar radiation (I_0) by a factor of 0·2 to allow for absorption of the longer and shorter wavelengths of light in the first meter of sea water. In shallow water columns this may be too large a factor and it is suggested instead that the incident solar radiation should be reduced by a factor of 0·5 to allow for the absorption

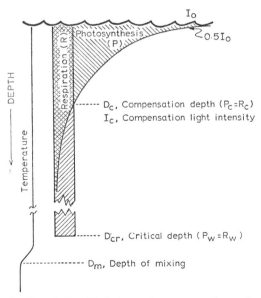

FIG. 31. Diagram showing the relationship between the compensation and critical depths, and the depth of mixing. (For explanation, see text).

of non-photosynthetic ultraviolet and infrared radiation in the first few centimeters of water. An average extinction coefficient for light (400 to 700 nm) penetrating the rest of the water column should then be used; the latter value may be rather larger than the value used by Sverdrup, especially if the model is used in coastal waters.

The model is illustrated in Fig. 31. If I_0 is the surface radiation and k is the average extinction coefficient, then the compensation light intensity (I_c) is related to the compensation depth (D_c) as

$$I_c = 0{\cdot}5 {\bullet} I_0 {\bullet} e^{-kD_c}. \tag{40}$$

At the compensation depth the photosynthesis of a cell (P_c) is equal to its respiration R_c; above this depth there is a net gain from photosynthesis ($P_c > R_c$) and below it there is a net loss ($P_c < R_c$). However, as the phytoplankton cells are mixed above and below the compensation depth they will experience an average light intensity ($\bar{I}$). The depth at which the average light intensity for the water column equals the compensation light intensity is known as the critical depth, D_{cr} [i.e. the depth at which photosynthesis for the water column (P_w) equals the respiration for the water column (R_w)]. The relationship of the critical depth to the compensation light intensity can be obtained by integrating eqn. (40) and dividing by the critical depth to obtain the average (compensation) light intensity for the water column. Thus

$$\bar{I}_c = 0{\cdot}5 {\cdot} I_0 \int_0^{D_{cr}} \frac{e^{-kD} {\cdot} dD}{D_{cr}}$$

$$\bar{I}_c = \frac{0{\cdot}5\, I_0}{kD_{cr}} (1 - e^{-kD_{cr}}). \tag{41}$$

From eqn. (41) it is possible to determine the critical depth for any water column, knowing the extinction coefficient (k), the solar radiation (I_0) and assuming some value for the compensation light intensity (see Section 3.1.2). Where D_{cr} is large the equation reduces to

$$D_{cr} = \frac{0{\cdot}5\, I_0}{\bar{I}_c k}. \tag{42}$$

From the explanation given above, it is apparent that if the critical depth is less than the depth of mixing, no net production can take place since $P_w < R_w$. This is illustrated as the condition in Fig. 31 where the depth of mixing (D_m), measured to the depth to the bottom of the principal thermocline, has been drawn as being greater than the critical depth. However, if the critical depth is greater than the depth of mixing, a net positive production will occur in the water column ($P_w > R_w$) and conditions for the onset of phytoplankton growth will have been established.

Several conditions are attached to the use of the critical depth model; these are (i) that plants are uniformly distributed in the mixed layer, (ii) there is no lack of plant nutrients, (iii) the extinction coefficient of the water column is constant (this in fact has to be determined as an average value), (iv) the production of the plants is proportional to the amount of radiation, and (v) respiration is constant with depth. The last of these assumptions is made as a matter of convenience and there is in fact no ecological data on the constancy of respiration with depth. Furthermore, some difficulties may be encountered in establishing the depth of mixing; in Fig. 31 this has been simplified as being the depth to the bottom of the principal thermocline. However, in other environments the depth of the principal halocline or the depth at which there is a maximum change of density with depth

$(\partial\varrho/\partial z \times 10^3$, Sverdrup *et al.*, 1946) may be better employed to determine the depth of mixing. A similar approach to the critical depth model was used by Riley (1946) and this is discussed in Section 4.2.2.

Once the mixing zone moves up above the critical depth, the standing stock of phytoplankton increases and a number of conditons for the critical-depth model may be no longer valid. Further as the biomass of phytoplankton increases, light conditions in the water column will be changed and the water column will become stratified. Empirical relationships between light penetration and chlorophyll *a* concentrations have been given by several authors (e.g. Riley, 1956; Aruga and Ichimura, 1968). In the former reference an equation is given relating the average extinction coefficient (k_e, m^{-1}) to the chlorophyll *a* concentration (C, mg Chl a/m^3) as follows:

$$k = 0 \cdot 04 + 0 \cdot 0088\,C + 0 \cdot 054 \cdot C^{2/3}. \tag{43}$$

Figure 32 shows diagrammatic examples of phytoplankton growth in a stratified water column. At time, t_0, just after active vertical mixing of water, phytoplankton is distributed

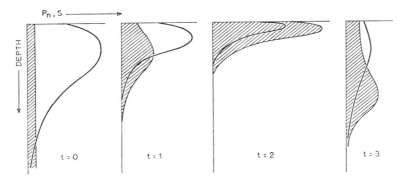

FIG. 32. Schematic changes in phytoplankton biomass (S) and daily net photosynthetic rate (P_n) after three time intervals (t) in stratified water (S or ▨, usually expressed in mg Chl$a$/m^3; P_n or —— usually expressed as mgC/mgChl$a$/day.

homogeneously in the water column but the *in situ* photosynthesis per unit phytoplankton biomass (mg C/mg Chl a/day) at each depth is different, being inhibited at the near surface by high light intensities and then decreasing from a subsurface maximum due to light attenuation. Assuming that the losses of phytoplankton at a given depth (from sinking, zooplankton grazing, etc.) are similar, the biomass of phytoplankton after a certain period can be estimated approximately by using eqn. (38) and (39). Because of the difference in the *in situ* photosynthesis per unit phytoplankton biomass at $t = 0$, the *in situ* growth rate of phytoplankton, μ will be different at each depth. Thus the profile of phytoplankton biomass and photosynthesis (called the 'productive structure'—Ichimura, 1956b) will change with time $t = 1$, $t = 2$, etc. As the phytoplankton biomass increases at the sub-surface photosynthetic maximum, the average extinction coefficient of light will also increase, and self-shading will occur (see Aruga, 1966). Accordingly, the daily compensation depth and the maximum in phytoplankton growth becomes shallower (Ichimura, 1956a). Providing there are sufficient nutrients, the pattern of photosynthesis and the standing stock of phytoplankton will finally maximize at the surface as a thin layer, and a bloom will occur ($t = 2$). As nutrients become exhausted in the surface layers, however, the depth of the maxima in the phytoplankton biomass and the primary productivity deepen ($t = 3$); the latter conditions occur in stratified water columns in temperate latitudes during the summer or may be found to prevail generally in tropical and sub-tropical waters throughout the year. However, other factors can modify the overall effect of light penetration on the production of the water column. Thus the feeding activity of zooplankton strongly affects the standing stock of phytoplankton and more detailed information of this subject is given in Section 4.2.2. In other cases, organic substances stimulate growth of phytoplankton (e.g. humic substances in sea water are reported to stimulate the growth of dinoflagellates—Prakash and Rashid, 1968). In some tropical areas motile phytoplankton, such as dinoflagellates (Eppley *et al.*, 1968), can use nutrients from the aphotic zone by diel migrations between near surface and deeper waters. A further method of overcoming nutrient limitation in tropical waters is through nitrogen fixation (e.g. Goering *et al.*, 1966), which sometimes gives rise to large surface blooms of *Trichodesmium*.

The quantity and quality of light in a water column is affected by the depth of the water column, the presence of dissolved coloured substances, and scattering due to suspended particles, including the phytoplankton. The light attenuation in a given water column is approximated from a logarithmic equation [eqn. (22) in Section 3.1.2] which is governed by the extinction, kd. The extinction is a function of the phytoplankton biomass (Ae), the depth of water (De) and the amount of other suspended and dissolved matter (Se), as follows (Sakamoto, 1966);

$$kd = \varepsilon\, Ae + \gamma\, De + \phi\, Se \tag{44}$$

in which ε, γ and ϕ are average extinction coefficients (400 to 700 nm) of a *unit* concentration or thickness of phytoplankton, water and other matter, respectively. Since it is considered that $\phi\, Se$ and $\gamma\, De$ (per unit thickness) are relatively constant in the euphotic zone, especially in pelagic waters, the variations in $\varepsilon\, Ae$ are primarily responsible for changes in the light environment. Assuming the daily compensation depth is the depth at which there is 1% of the surface illumination,* then the relation between chlorophyll a in the euphotic zone (Ae, mg Chl a/m^2), the depth of euphotic zone (De, in meters) and suspended and dissolved matter (Se, expressed in units of thickness, e.g. meters) can be expressed by the following equation:

$$0{\cdot}01 = e^{-(\varepsilon\, Ae + \gamma\, De + \phi\, Se)} \tag{45}$$

$0{\cdot}01$ can be approximately replaced by $e^{-4{\cdot}6}$, which gives:

$$4{\cdot}6 = \varepsilon\, Ae + \gamma\, De + \phi\, Se \tag{46}$$

then
$$Ae = \frac{4{\cdot}6 - (\gamma De + \phi Se)}{\varepsilon}, \tag{47}$$

where Ae in eqn. (47) represents the total chlorophyll a in the euphotic zone of a given water column. If the euphotic depth is close to zero (i.e. the light absorption due to water and suspended and dissolved matter is negligible, $\gamma\, De + \phi\, Se \simeq 0$), $Ae_{\max} \simeq 4{\cdot}6/\varepsilon$; where

---

* Compensation light intensities for marine phytoplankton are generally found to be in the approximate range of $0{\cdot}002$ to $0{\cdot}009$ ly/min; since these values are very approximately two orders of magnitude lower than average surface radiation values, the compensation depth is sometimes simply estimated from the depth at which there is 1% of the surface radiation. A further approximate estimate of this depth is based on the relationship between the depth of the Secchi disc and the extinction coefficient. If this is assumed to be $k_e = 2{\cdot}1/T$ (p. 78.), then $0{\cdot}01 = e^{-kDc}$ and $kDc = 4{\cdot}6$ substituting for k with the Secchi depth (T), $Dc = (4{\cdot}6/2{\cdot}1)$. $T \simeq 2T$. Thus twice the Secchi depth is sometimes employed as an estimate of the euphotic depth.

TABLE 20. AVERAGE LIGHT EXTINCTION COEFFICIENTS (400 TO 700 nm) OF CULTURED ALGAE AND SEA WATER (FROM TAKAHASHI AND PARSONS, 1972).

Materials	ε (mg Chl a/m^2)$^{-1}$	Chl a·max (mg Chl a·m^{-2})	References
Chlorella sp.	0·0115	400	Tominaga & Ichimura (1966)
Scenedesmus sp.	0·0051	900	Aruga (1966)
Cyclotella meneghiniana	0·0048	960	Steemann Nielsen (1962)
Skeletonema costatum	0·0071	650	Tominaga & Ichimura (1966)
Natural phytoplankton populations (including suspended particles and dissolved matter, Kuroshio)	0·0184	250	Aruga & Ichimura (1968)
Sea water (clear ocean)	$\dfrac{\phi+\gamma\;(\text{m}^{-1})}{0\cdot045}$		Jerlov (1957)

Ae_{max} represents the maximum amount of chlorophyll a in the euphotic zone of a given water column. Some values for the maximum amount of chlorophyll a in cultures are shown in Table 20. *Skeletonema* culture shows a maximum of 650 mg Chl a/m^2. Under field conditions, the maximum amount of chlorophyll a is usually less than that of pure cultures because of the effects of ϕ Se and sometimes γ De on the increase of kd. For example, in the Kuroshio current, the maximum amount of chlorophyll a in the euphotic zone has been computed to be 250 mg Chl a/m^2. Fig. 33 shows the relationship between total amount of

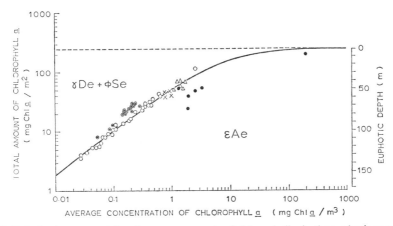

FIG. 33. Relation between total and average amounts of chlorophyll a in the euphotic zone in the Pacific Ocean. (● Bays; △ Oyashio; ○ Kuroshio; × mixed region of the Oyashio and Kuroshio; ⊙ Pacific Ocean, 155°W, 50°N–10°S; from Takahashi and Parsons, 1972).

chlorophyll a/m^2 and the average concentration of chlorophyll a in the euphotic zone in the Pacific Ocean. In pelagic waters, at least in the Pacific Ocean, the effects of dissolved matter and suspended particles (other than phytoplankton) on light attenuation seem to be uniform in the euphotic zone. Consequently, in Fig. 33, the decrease in average concentration of chlorophyll a causes an increase in the euphotic depth since kd is a function of the total amount of chlorophyll a. If the maximum amount of chlorophyll a is assumed to be 650 mg Chl a/m^2 (e.g. *Skeletonema* in Table 20), the discrepancy between the maximum and total chlorophyll a will be a function of $\phi\ Se + \gamma\ De$, which increases with increasing euphotic depth. Thus the maximum productivity of deep euphotic water columns must be less than shallow euphotic water columns, based on the maximum standing stock of chlorophyll a.

$\varepsilon\ Ae$ as discussed here was used as an indicator of energy efficiency in photosynthetic production (Platt, 1969). Platt found that $\varepsilon\ Ae$ (k_b in his text) was approximated by the ratio of photosynthesis ($P_{(w)}$) to illumination ($I_{(0)}$), when both were expressed in calories; $\varepsilon\ Ae \simeq P_{(w)}/I_{(0)}$. Consequently, larger values of $\varepsilon\ Ae$ cause higher photosynthetic efficiency.

It is predictable that the daily photosynthesis in a given water column will change with sustained changes in the radiant energy reaching the sea surface, even if the amount of chlorophyll a does not change. As an example, differences in daily net photosynthesis under different amounts of solar radiation are shown in Table 21. For the calculation of results

TABLE 21. DAILY NET PHOTOSYNTHESIS FOR VARIOUS AMOUNTS OF CHLOROPHYLL a UNDER DIFFERENT LEVELS OF IRRADIANCE (FROM TAKAHASHI AND PARSONS, 1972).

Radiation (ly/day)	Euphotic depth (M)	Net photosynthesis (mg C/m²/day)								
		175	150	100	50	30	20	10	5	<1
	Chl a amount (mg/m²)	1	2	10	52	100	135	190	220	260
100		1·6	3	16	83	160	220	300	350	420
200		2·6	5	26	140	260	350	490	570	680
300		3·3	7	33	170	330	450	630	740	860
400		4·0	8	40	210	400	540	760	880	1000
500		4·7	9	47	240	470	630	890	1000	1200
600		5·3	11	53	280	530	720	1000	1200	1400
700		5·8	12	58	300	580	780	1100	1300	1500
800		6·3	13	63	330	630	850	1200	1400	1600
900		6·8	14	68	350	680	920	1300	1500	1800

shown in this table, the photosynthesis vs light curve shown in Fig. 26B (in Section 3.1.2) was used, the P_{max} was assumed to be 1 mg C/mg Chl a/hr and the relation between the euphotic depth and the amount of chlorophyll a in the euphotic zone was obtained from Fig. 33. If data on the total surface radiation and the total amount of chlorophyll a in the euphotic zone (or the depth of the euphotic zone) can be obtained, one can estimate the approximate net photosynthesis from Table 21. In calculating production in Table 21, P_{max} was assumed to be 1 mg C/mg Chl a/hr, which falls in the range of values for temperate pelagic waters (see Table 17). It would be better to measure the actual P_{max} in any environment in order to obtain a better prediction of the *in situ* productivity; in such a case the values for net photosynthesis in Table 21 have to be multiplied by the value of the actual P_{max}. In highly productive areas, P_{max} may be greater than 5 mg C/mg Chl a/hr. Thus the calculated net photosynthesis from Table 21 could be at least $1800 \times 5 = 9000$ mg C/m²/day. This value is in general agreement with maximum *in situ* primary productivity values of 5 to 10 g C/m²/day. Primary productivity values under culture conditions may be two or three times greater than *in situ* (see also Ryther, 1963).

3.1.5 THE KINETICS OF NUTRIENT UPTAKE

The two earliest ecological considerations in the uptake of nutrients by phytoplankton were firstly, that at low nutrient concentrations, the rate of nutrient uptake was found to be concentration dependent, and secondly that the total yield of phytoplankton was directly proportional to the initial concentration of limiting nutrient and independent of the growth rate of the phytoplankton. Ketchum (1939a) demonstrated the first of these results; using the diatom, *Nitzschia closterium*, he showed that the uptake of nitrate was concentration dependent over an approximate range from 1 to 7 µg at N/l. Spencer (1954), using the same organisms, demonstrated the second result when he showed a linear relationship between the initial concentration of nitrate and total yield of cells over nitrate concentrations ranging from approximately 15 to 150 µg at/l. Ketchum's results also showed that another nutrient, phosphate, followed similar kinetics to the uptake kinetics of nitrate and that phosphate could be taken up in the absence of measurable nitrate; however, phosphate uptake was increased when the concentration of nitrate was increased.

Taking the original work of Ketchum a step further, Caperon (1967) and Dugdale (1967) showed that nutrient uptake could be described using Michaelis–Menten enzyme kinetics in which

$$v = \frac{V_m S}{K_s + S},\tag{48}$$

where v is the rate of nutrient uptake, V_m is the maximum rate of nutrient uptake, K_s is the substrate concentration at which $v = V_m/2$ and S is the concentration of nutrient. By plotting S/v versus S, a straight line is obtained with an intercept on the abscissa $-K_s$. The constant K_s is believed to be an important property of a phytoplankton cell since it reflects the ability of a species to take up low concentrations of a nutrient and as such it may determine the minimum nutrient concentration at which a species can grow. Thus MacIsaac

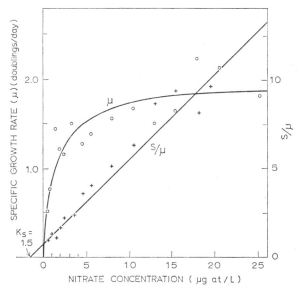

FIG. 34. Specific growth rate of *Asterionella japonica* as a function of nitrate concentration (redrawn from Eppley and Thomas, 1969).

and Dugdale (1969) have shown that coastal phytoplankton communities generally have K_s value of >1 µM for nitrate uptake, while oceanic communities have lower K_s values of ca 0·2 µM.* In line with other biochemical studies on Michaelis–Menten constants it is also important to recognize that K_s values are temperature dependent (Eppley et al., 1969b).

An important question to resolve is whether the K_s of a phytoplankton species in the sea can be determined only from the rate of nutrient uptake or whether the same value can be determined from the phytoplankton growth rate. It has been demonstrated experimentally that nutrient uptake can occur without cell division (Fitzgerald, 1968); ecologically, however, it is important to know whether this is a general phenomenon. From studies conducted by Eppley and Thomas (1969) on two species of diatom it was concluded that the half-saturation constants for growth and nitrate uptake were very similar so that the expression given above for the velocity of nutrient uptake can also be given in the form

$$\mu = \mu_{max} \frac{S}{K_s + S},$$ (49)

TABLE 22. HALF-SATURATION CONSTANTS FOR NITRATE AND AMMONIA UPTAKE.

Phytoplankton species, clone or area	K_s (µgat/l) Nitrate	K_s (µgat/l) Ammonia	Reference
Oligotrophic, tropical Pacific	0·04 0·21 0·01 0·03 0·14	0·10 0·55 0·62	MacIsaac and Dugdale (1969)
Eutrophic, tropical Pacific	0·98		
Eutrophic, subartcic Pacific	4·21	1·30	
Oceanic species	0·1 to 0·7	0·1 to 0·4	Eppley et al., (1969b)
Neritic diatoms	0·4 to 5·1	0·5 to 9·3	
Neritic or littoral flagellates	0·1 to 10·3	0·1 to 5·7	
Cyclotella nana Clone 3H Clone 7–15 Clone 13–1	1·87 1·19 0·38		Carpenter and Guillard (1971)
Fragilaria pinnata Clone 13–3 Clone 0–12	0·62 1·64		

* 0·2 µM nitrate $\equiv$ 0·2 µgat NO_3^-.

where μ and μ_{max} are the growth rate and maximum growth rate, respectively, in units of time^{-1}. A plot of the above expression is given in Fig. 34 together with a plot of S/μ vs S, which gives the intercept value on the abscissa of K_s. A summary of some K_s values are given in Table 22. From these data it was suggested by Eppley et al., (1969b) that K_s varied approximately with cell size; thus Coccolithus huxleyi, which is ca 5 μ diameter, had a K_s of 0·1 μgat NO_3^-/l while Gonyaulax polyedra, which is ca 45 μ diameter, had a K_s of >5 μgat NO_3^-/l. It was also apparent from additional experiments conducted by Eppley et al., (1969b) that the ability to take up nitrate and ammonium ions differed between species and that some species could take up nitrate at lower concentrations than they could take up ammonia, and vice versa. Carpenter and Guillard (1971) showed that differences in K_s values were not confined to species but that clones of the same species had different K_s values depending on their environment. Clones isolated from low nutrient oceanic water had K_s values of <0·75 μgat/l while the same species taken from an estuarine region had a K_s of >1·5 μgat/l.

Differences in K_s values coupled with differences in the ability of species to reach their maximum growth rates at different light intensities, have been suggested by Dugdale (1967) and Eppley et al., (1969b) as being important factors in determining species succession in phytoplankton blooms. A schematic example is given in Fig. 35; in the first figure it may

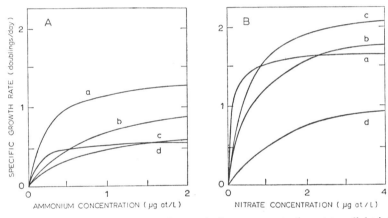

FIG. 35. Specific growth rate vs. ammonium and nitrate concentration at two light intensities, (B) approx. 4 times (A). (a) Coccolithus huxleyi, (b) Ditylum brightwelli, (c) Skeletonema costatum, (d) Dunaliella tertiolecta (redrawn from Eppley et al., 1969 b).

be seen that Coccolithus huxleyi reached its maximum growth at a low light intensity using ammonium ions and that its growth was approximately double that of the other species at a concentration of 0·5 μgat NH_4^+/l. However, in the second figure, at a higher light intensity, both Skeletonema costatum and Ditylum brightwellii grew faster than C. huxleyi at concentrations above 2 μgat NO_3^-/l. Since specific growth rates are exponents it is quite apparent that even small differences, over a period of time, would lead to very large differences in the standing stock of a particular species. The validity of this approach to field studies has been demonstrated by Thomas (1970b) and Thomas and Owen (1971), who showed that it was possible to obtain good agreement between growth rates calculated from $^{14}CO_2$ uptake and from ammonium concentrations in tropical Pacific waters. However, this empirical justification for the determination of K_s and phytoplankton growth constants from field data is not wholly supported by experimental and theoretical considerations advanced by several authors. The relationship between growth rate and nutrient concentration described in the above equation depends on μ_{max} and K_s remaining constant; this condition is generally met during exponential growth when the nutrient content per cell remains relatively constant, such that

$$\frac{\text{uptake rate}}{\text{cell division rate}} = Q, \text{ the nutrient content per cell.} \tag{50}$$

However, Kuenzler and Ketchum (1962) showed that the uptake of phosphorus by *Phaeodactylum* was not determined by the phosphate concentration in the medium but by the past history of the cells and their internal phosphorus content. The cell content of phosphorus varied between 2 and 66×10^{-15} moles/cell; Caperon (1968) showed a similar variability in the nitrogen content of *Isochrysis galbana*, from 2 to 40×10^{-15} moles/cell. Further it has been known from some of the earliest studies (e.g. Ketchum, 1939a) that nutrient uptake can take place in the dark when the cell division rate is zero. Eppley and Strickland (1968) have suggested that phytoplankton cells require a certain minimum content of nitrogen (or phosphorus) before cell division can proceed; for *Dunaliella tertiolecta* the minimal nitrogen per cell (Q_0) varied with light intensity between 0.7×10^{-15} and 1.4×10^{-15} moles/cell. Caperon (1968) showed that there was a hyperbolic relationship between the reserve nitrate content per cell (Q') and growth rate (μ), such that

$$\mu = \mu_{max} \cdot \frac{Q'}{A + Q'}, \tag{51}$$

where A is a growth constant. The value Q' used in the above equation represents the nitrate in the cell reservoir which is in addition to Q_0, the minimum amount of nitrate required for a normal cell at zero growth rate. Thus A represents the concentration of reservoir nitrate within the cell which is necessary to obtain a growth rate of $\mu_{max}/2$. In chemostat studies on vitamin B_{12} requirements, Droop (1970) showed that the growth rate could be expressed in terms of Q_0 and Q as

$$\mu = \mu_{max} \left(1 - \frac{Q_0}{Q} \right). \tag{52}$$

From these findings it is apparent that the ecological use of K_s values (Table 21) should be approached with caution. As pointed out by Eppley *et al.* (1969b), for example, the dinoflagellate, *Gonyaulax polyedra*, has a low growth rate and an apparently high K_s. However, this organism displays a diel migration in which it absorbs nutrients from 10 to 15 m during the night and swims to the surface during the day where light is available for photosynthesis and growth, but where the nutrients are much lower than at 10 to 15 m. Thus the uptake of nutrients can be independent of growth and the phenomenon of "luxury consumption" (see Eppley and Strickland, 1968) may result in growth being proportional to the nutrient content per cell (e.g. Droop, 1968 and 1970) rather than the nutrient concentration *in situ* (e.g. Thomas, 1970b).

Nutrient-dependent differences in growth rates may account for the dominance of one species over another on the basis of their respective growth rates but as Hulbert (1970) has pointed out, there is little or no possibility of an abundant species monopolizing the nutrient supply and forcing a less abundant species to extinction. Each algal cell in a phytoplankton community may be visualized as the center of a volume of water in which nutrient depletion decreases from the cell outwards; this volume may be referred to as the cell's nutrient zone.

Hulbert (1970) based his conclusions on the fact that the nutrient zone surrounding a cell would have to overlap with that of another cell if one growth rate was to affect another. For cellular nutrient zones to overlap, Hulbert calculated that there would have to be at least 3×10^8 phytoplankton cells per litre; since most natural populations rarely exceed 10^6 cells/litre in open and coastal waters, there is generally no possibility of one nutrient zone overlapping with another. However, at cell concentrations of 10^9/litre, such as may occur in some estuarine and eutrophic environments, there is a possibility that dominant species will lead to the elimination of residual species and hence decrease the diversity of the system.

3.1.6 CHEMOSYNTHESIS

Some micro-organisms can satisfy their primary energy requirements by utilizing simple inorganic compounds, such as ammonia, methane, or nitrite, or elements, such as ferrous iron, hydrogen gas, or water insoluble amorphous sulfur. All of the known organisms which comprise this (chemosynthetic) group are bacteria. Most of their inorganic substrates are derived from the decay of organic matter which is itself primarily formed through photosynthesis. Consequently, if one follows the origin of the energy used in chemosynthesis, chemosynthetic processes may not be considered as primary production. However, chemosynthesis usually involves carbon dioxide fixation and the primary formation of new particulate material. Thus chemosynthesis may be considered as a special case of primary production on the grounds of its trophic position in the marine food chain.

In the past it was believed that the inorganic substrates reacted directly with molecular oxygen to form oxidized end-products. However, Bunker (1936) showed that molecular oxygen was not directly involved in sulfur oxidation by thiobacilli. The fact that oxygen in the end-products originated from water resulted in a new concept of chemosynthesis, although in some cases it is still questionable whether the concept is adaptable to all cases of chemosynthetic activity. Thus reactions of the type suggested by Bunker involve dehydrogenations rather than oxidations. In general terms the chemosynthetic process can be expressed conveniently in three stages according to Bunker's concept (Gundersen, 1968).

(1) As the result of dehydrogenation, high reducing power is produced as follows:

$$n\text{AH}_2 + n\text{H}_2\text{O} \xrightarrow{\text{dehydrogenase}} n\text{AO} + 4n[\text{H}^+ + \text{e}^-] \tag{53}$$

$$\underset{\text{(Inorganic substrates)}}{} \quad \underset{\substack{\text{(oxidized} \\ \text{end-product)}}}{} \quad \underset{\text{(reducing power)}}{}$$

where the symbol $[\text{H}^+ + \text{e}^-]$ is used as a synonym for the reducing power.

(2) A proportion of the reducing power is then utilized for energy production (adenosine triphosphate, ATP, synthesis) by being transferred through the cytochrome system to molecular oxygen. A second part of the reducing power is transferred to NAD (nicotinamide adenine dinucleotide) in order to produce reduced NAD (or NADH$_2$, reduced nicotinamide adenine dinucleotide). These relationships might be visualized as follows:

$$4[\text{H}^+ + \text{e}^-] + m\text{ADP} + m\text{P}_i + \text{O}_2 \xrightarrow[\text{system}]{\text{cytochrome}} 2\text{H}_2\text{O} + m\text{ATP} \tag{54}$$

$$2[\text{H}^+ + \text{e}^-] + \text{NAD} \longrightarrow \text{NADH}_2 \tag{55}$$

where ADP is adenosine diphosphate, and P_i is inorganic phosphate. Some anaerobes can use bound oxygen derived from inorganic compounds instead of free oxygen as shown in the above equation (e.g. sulphate-reducing bacteria).

(3) The ATP and $NADH_2$ are then used for the assimilation of carbon dioxide:

$$12\, NADH_2 + 18\, ATP + 6\, CO_2 \rightarrow C_6H_{12}O_6 + 6\, H_2O + 18\, ADP + 18\, P_i + 12\, NAD \quad (56)$$

Thus the different inorganic substrates used by chemosynthetic bacteria are not merely the sole source of the organism's energy but also the sole source of their reducing power.

Depending on differences in the inorganic substrate (AH_2), chemosynthetic bacteria are classified into several groups, such as nitrifying, sulfur, hydrogen, methane, iron, and carbon monoxide bacteria. Table 23 shows some representative chemosynthetic bacteria inhabiting marine environments. The overall efficiencies of chemosynthesis during the growth of the bacteria (e.g. nitrifying or sulfur bacteria) can be expressed as the ratio between the total energy consumed in carbon dioxide assimilation and the energy liberated by the primary inorganic compounds during oxidation; these efficiencies are 6 to 8% (Gibbs and Schiff, 1960) although the efficiency may sometimes change drastically with different stages of cultures. For example, young cultures of *Nitrosomonas* gave values approaching 50%; in older cultures the efficiency dropped to 7% (Hofmann and Lees, 1952). Most chemosynthetic bacteria require free oxygen as the electron acceptor in the second step described above. However, facultative or obligate anaerobic bacteria, such as *Thiobacillus denitrificans* and *Desulfovibrio desulfricans*, can use bound oxygen derived from nitrate or sulfate.

Among inorganic substrates available for the chemosynthetic bacteria, nitrogen and sulfur compounds are relatively abundant and widely distributed compared with the other reduced compounds in the pelagic environment of the oceans. Among reduced nitrogen compounds, ammonia may be present in concentrations up to *ca* 5 μg at/l and nitrite at concentrations up to *ca* 2 μg at/l (Riley and Chester, 1971). Actual occurrences of nitrifying bacteria, expressed in colonies per litre of sea water, were found to be <1 in the north Atlantic Ocean, <10 in the Pacific Ocean, *ca* 10^4 in the Indian Ocean near an island, and *ca* 10^6 in Barbados harbor (Watson, 1965; Hattori and Wada, 1971). Some of the strains of nitrifying bacteria isolated from marine waters are the same or similar to those from fresh water or soil, but others are peculiar to marine environments (e.g., see Watson, 1971). Nitrification by marine bacteria, using either ammonia or nitrite as a substrate, is reported to be more efficient at low (<0·1 ml/l) oxygen concentrations (Carlucci and McNally, 1969).

The quantities of reduced sulfur compounds in the marine environment are generally much greater than the quantities of reduced nitrogen compounds. Thiosulfate and polythionates may sometimes be present at levels from 0 to 100 μg at/l, the highest concentrations of these compounds being detected near to shore (Tilton, 1968). Sulphides are generally not detected within the analytical limit of <1 μg at/l in open ocean waters (Tilton, 1968). The number of colonies of sulfur bacteria (*Thiobacillus* spp.) have been found to range from 0 to 275 per 100 ml (Tilton *et al.*, 1967). These values for colony counts are several orders of magnitude smaller than those (10^3 to 10^4 per 100 ml) which were calculated by Tilton *et al.* (1967) on the basis of the amount of reduced sulfur compounds.

The reduced inorganic substrates are mainly produced through anaerobic metabolic processes. Consequently, anaerobic environments, which sometimes develop in fjords and estuaries, create favourable habitats for chemosynthetic bacteria. Under such conditions vigorous growth of certain species of chemosynthetic bacteria (usually sulfur, hydrogen, or methane bacteria) has been observed (Kuznetsov, 1959; Sokolova and Karavaiko, 1964).

It should be mentioned that an obligate anaerobic strain, *Desulfovibrio desulfricans*, has been isolated even from oxic ocean waters (Kimata *et al.*, 1955; Wood, 1958). In order to explain this phenomenon, Baas Becking and Wood (1955) proposed the idea of 'metabiosis'

TABLE 23. SUMMARY OF CHEMOSYNTHETIC BACTERIA GROWING IN THE OCEAN AND THEIR INORGANIC SUBSTRATES.

	Inorganic substrate	Oxidized end-product	Oxidizer	ΔF^* (Kcal)	Ability to grow heterotrophically	Habitat
(1) Nitrifying bacteria						
Nitrobacter spp.	NO_2	NO_3	O_2	18		Soil, fresh, and sea waters
Nitrococcus mobilis WATSON & WATERBURG	NO_2	NO_3	O_2	18		Sea water
Nitrospina gracilis WATSON & WATERBURG	NO_2	NO_3	O_2	18		Sea water
Nitrosomonas spp.	NH_3	NO_2	O_2	85		Soil, fresh, and sea waters
Nitrosococcus oceanus (WATSON) comb. nov.	NH_3 or NH_2OH	NO_2	O_2	85		Sea water
(2) Sulfur bacteria						
Thiovulum majas HINZE	H_2S	S	O_2	50	?	Sea water
Beggiatoa spp.	H_2S	S	O_2	50	+	Fresh and sea waters
Thiospira bipunctata MOLISCH	H_2S	S	O_2	50	?	Sea water
Thiothrix spp.	S	SO_4^{-}	O_2	119	?	Soil, fresh, and sea waters
Thiobacillus thioparus BEIJERINCK	$5/4\ S_2O_3^{-}$	$3/2\ SO_4^{--}+s$	O_2	112	+	Soil, fresh, and sea waters
Thiobacillus denitrificans BEIJERINCK	$5S$	$5SO_4^{--}$	O_2	112	+	Soil, fresh, and sea waters
(3) Hydrogen bacteria						
Hydrogenomonas spp.	H_2	H_2O	O_2	56	+	Soil, fresh, and sea waters
Desulfovibrio desulfricans (BEIJERINCK) KLUYVER & VAN NIEL	H_2	H_2O	SO_4	56	+	Soil, fresh, and sea waters
(4) Methane bacteria						
Methanomonas spp.	CH_4	CO_2	O_2		+	Soil, fresh, and sea waters
(5) Iron bacteria						
Grallionella spp.	$4FeCO_3$	$4Fe(OH)_3$	O_2	81	+	Soil, fresh, and sea waters
(6) Carbon monoxide bacteria						
Sarcina bakerii SCHNELLEN	CO	CH_4	H_2		+	Soil, fresh, and sea waters

*Free energies per number of moles of electron donor indicated. Values reviewed by Gibbs and Schiff (1960) are mainly quoted here.

in which they suggested that several kinds of microorganisms co-exist on or in bacterial aggregates and conditions for the production of reduced substrates could occur within the aggregates. Thus the obligate anaerobic bacteria existing in oxic environments are probably present within the microcosm of a bacterial aggregate (Seki, 1972).

In the ocean, chemosynthetic activity can be estimated from carbon dioxide uptake in the dark using $NaH^{14}CO_3$. Compared with carbon fixation through photosynthesis, dark CO_2 uptake is usually small (i.e. less than 5% of the photosynthesis on a daily basis within the euphotic zone in pelagic areas). Dark CO_2 uptake is not entirely carried out by chemosynthesis, but heterotrophic processes by bacteria and algae (e.g. Wood and Werkman, 1935, 1936 and 1940) may result in the uptake of small amounts of CO_2 in the absence of light. Algal dependency on dark fixation of CO_2 is low (i.e. 3 to 5% of the total CO_2 fixed in the light) and this is about the same proportion of CO_2 fixed by aerobic heterotrophic bacteria when they are growing on an organic substrate (Sorokin, 1961 and 1966). However, among facultative autotrophic bacteria which belong to an intermediate metabolic type between obligate heterotrophs and chemoautotrophs, there is a requirement of between 20 and 90% CO_2 during the oxidation of low molecular weight organic compounds, such as methane and formic acid; for obligate chemoautotrophs the requirement for CO_2 is close to 100%. Since in natural environments, microorganisms which depend on different kinds of substrates exist together, it is practically impossible to separate out and estimate the actual activity of chemosynthetic organisms except at the time of vigorous growth of any one species. However, a measure of dark CO_2 uptake is a useful measure of chemosynthetic activity in any environment. As an example, Seki (1968) studied seasonal changes in dark uptake of CO_2 in a small bay and showed that dark uptake of CO_2 near the bottom of the bay could be as high as 20% of the photosynthesis throughout the year and that during the spring dark uptake for the water column was sometimes 50% greater than photosynthesis. If B_t is the bacterial biomass produced (expressed in mg C/l/day) and B_i (expressed in mg C/l/day) denotes the inorganic carbon incorporated, then $B_i = f \cdot B_t$ (Romanenko 1964a, b) where f is the coefficient of inorganic carbon dependency of microorganisms. B_t can be measured directly as the dark uptake of $^{14}CO_2$. B_t can be determined microscopically and the increase in bacterial cell numbers measured during an incubation period (usually one day). B_t is then converted to mg C/l/day using assumed factors of 0·15 for wet weight of bacteria to dry weight, and 0·5 for dry weight to carbon. Very approximate values of the factor, f, can be calculated for a number of different organisms; Romanenko (1964c) and Sorokin (1971) have reported the following average factors for obligate chemoautotrophs, facultative autotrophs and heterotrophic bacteria.

Metabolic Process	Substrate (Examples)	% CO_2 Required in Biosynthesis	f
Aerobic	Glucose	4	0·04
Mostly facultative anaerobic	Methane, Formic acid	25	0·25
Mostly anaerobic	Ammonia, Hydrogen sulphide	100	1·00

Active chemosynthesis occurs in waters and sediments in which both aerobic and anaerobic environments exist in the same column. Sorokin (1964a and b) has studied the importance of chemosynthetic bacteria in the Black Sea, where a thick anaerobic zone exists below the depths of *ca* 150 m in the central part of the sea. A summary of his finding is shown in Fig. 36. At depths shallower than 50 m, inorganic carbon is fixed mostly through photosynthesis by algae, and the daily photosynthesis is reported to be *ca* 350 mg $C/m^2/day$ during October. At the transition zone between 100 and 250 m, where environmental conditions are changing from aerobic to anaerobic, chemosynthetic fixation of inorganic carbon is predominant. An *in situ* maximum value for daily chemosynthesis was *ca* 9 mg

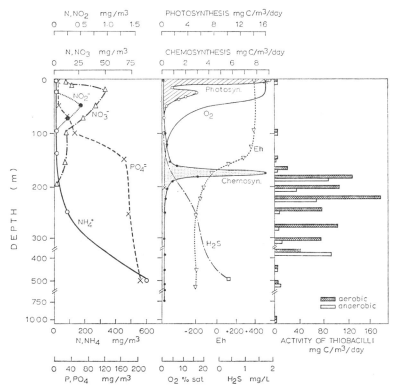

FIG. 36. Daily chemosynthetic production and some factors influencing it in the Black Sea (after Sorokin, 1964 c).

$C/m^3/day$; chemosynthesis for the water column amounted to approximately 200 mg $C/m^2/day$. This active chemosynthesis is carried out mainly by sulfur bacteria (aerobic and anaerobic *Thiobacillus*), and it is supported by a continual supply of reduced sulfur compounds, such as H_2S, S, and S_2O_3, from the anaerobic zone. The potential activity of thiobacilli in water samples taken from the depth of their maximum activity is very high compared with the *in situ* activity; by determining the oxidation rate of thiosulfate added to a water sample, a potential activity of 200 mg $C/m^3/day$ has been obtained.

Although the studies described above may appear to have limited geographical significance, it is probable in fact that similar microzonation in environmental factors can occur in many marine environments, especially in bays and estuaries. Also in all near-shore sediments where wave action does not disturb the benthos, a high chemosynthetic activity will occur in a depth zone of a few cm just below the sediment surface. If the transient zone of aerobic to anaerobic conditions comes up above the depth receiving a few per cent of the surface illumination, anaerobic photosynthetic sulfur bacteria, which require reductants (e.g. H_2S, S) for the hydrogen donor, grow vigorously. These organisms impart a purple or deep green colour to the sediment or water column in which they exist.

3.2 HETEROTROPHIC PROCESSES

3.2.1 THE ORIGIN OF ORGANIC SUBSTRATES

Heterotrophic organisms are dependent on an organic carbon source to provide energy for growth. In the sea, most of the organic carbon utilized by heterotrophic organisms originates from the marine biota, and only in near-shore coastal areas is there any appreciable contribution of organic materials from the land. Exceptions to this generalization are found where major rivers may influence the oceanic environment for a considerable distance off shore; in particular, the Amazon River, with an annual discharge of 6×10^{12} m³ (or *ca* 20% of the entire world wide river runoff), may contain sufficient organic carbon (2 to 10 mg/l) to influence heterotrophic activity over an oceanic area approximately 10^7 km² during maximum runoff (Williams, 1968). Duursma (1965) recognized four main groups of organic compounds which occur as dissolved substances in sea water. These are (1) nitrogen-free organic matter, including carbohydrates, (2) nitrogenous substances, including amino acids and peptides, (3) fat-like substances, and (4) complex substances, including humic acids, derived from groups (1) and (2). In addition to these dissolved materials, particulate organic materials also serve as a substrate for heterotrophic organisms. The nature and chemical composition of these substances are discussed in Section 1.4; their origin is illustrated in Fig. 37, which is adapted from Johannes (1968). The figure illustrates

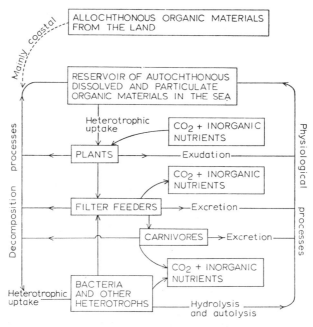

Fig. 37. Pathways of transfer and regeneration of organic substrates in an aquatic ecosystem (modified from Johannes, 1968).

that the two principal pathways of organic materials which act as substrates for heterotrophic organisms result from the food chain of the sea, firstly through the physiological release of materials and secondly through the decomposition of plants and animals themselves. Autochthonous organic materials derived by physiological processes include the release of dissolved organic materials by phytoplankton; this subject has been reviewed by Fogg (1966), who records that in some cases, up to 50% of the CO_2 photosynthetically fixed by phytoplankton may be released as soluble organic carbon (Allen, 1956; Fogg, 1952). This process is sometimes called 'excretion' but should probably be called 'exudation'. The latter term is used by Sieburth (1969) and Sieburth and Jensen (1969) in referring to the loss of soluble organic material from seaweeds; the authors report that this loss may amount to 40% of the net carbon fixed daily. Further studies on the exudation of soluble organic carbon by phytoplankton (Eppley and Sloan, 1965; Hellebust, 1965) have shown that in general the release from healthy cells amount to 15% or less of the total carbon fixed; Watt (1966) has suggested a maximum loss of extracellular products of up to 30%. Two factors which appear to effect the release of soluble organics are the age of the cells and the light intensity. Hellebust (1965) and Anderson and Zeutschel (1970) have shown that exudation is greatest at high light intensities and among phytoplankton cells collected at the end of a plankton bloom.

It is quite apparent from these studies that since plants are the major producers of organic matter in the sea, any fraction lost through exudation will constitute an appreciable input to the organic reservoir in the environment. The type of compounds released by algae are known to include small amounts of amino acids and generally larger amounts of short chain acids (e.g. glycollic acid), glycerol, carbohydrates, and polysaccharides (Fogg, 1966; Watt, 1966).

The excretion of organic materials resulting from digestive processes of animals forms a second physiological pathway for the input of organic materials available for heterotrophic growth. In addition to the process of true excretion, animals may also release organic material directly from their bodies (e.g. Johannes and Webb, 1970), and this may in part be coupled with the moulting process, which results in the specific production of chitinous debris; for example, Lasker (1966) estimated that the loss of chitin by euphausiids amounted to 10% of their body weight every five days.

The natural mortality of plants and animals in the food chain forms the main pathway for the production of organic substrates by decomposition processes. Post-mortem changes in the permeability of cell membranes and the effect of autolytic enzymes (Krause et al., 1961; Krause, 1961 and 1962) may result in an initial loss of 15 to 50% of the total biomass of an organism. Soluble organic compounds released from dead organisms include amino acids, peptides, carbohydrates, and fatty acids.

Heterotrophs which utilize the organic particulate and soluble substrates produced by the food chain, themselves contribute to the organic reservoir of the sea, either by serving as food for filter feeding organisms or by hydrolysing substrates, and self autolysis, to provide soluble organic materials. Williams (1970) found that soluble organic materials were assimilated by bacteria with an average efficiency of 67% for glucose and 78% for amino acids. Sorokin (1970) found that algal hydrolysates were utilized by natural populations of marine heterotrophs with an efficiency of ca 45%. These efficiencies are much higher than are found in culture experiments (e.g. Parsons and Seki, 1970) where only about one third of the organic substrate taken up is retained as new cellular material. Williams (1970) suggests that higher conversion efficiencies result with natural populations utilizing a variety of organic compounds, in contrast to the usual laboratory studies, where all organic carbon must be derived from a single organic compound. The largest fraction of organic carbon lost during heterotrophic activity is respired back to carbon dioxide, a process which also occurs throughout the food chain as organic carbon is transferred to higher trophic levels.

While information summarized in Fig. 37 shows the major pathways for the release of organic compounds in the heterotrophic cycle of the sea, there are in addition very specific organic compounds released in sea water which can modify the overall balance of organic materials in any one environment. Included among specific organic substances are growth factors and antimetabolites. As an example of the former, Burkholder and Burkholder (1956) and Starr (1956) showed that in coastal areas, vitamin B_{12} is produced by the bacteria; secretion of vitamins by some phytoplankton has also been demonstrated (Carlucci and Bowes, 1970). Since other species of phytoplankton require vitamins for growth, their presence can be assumed to affect the organic food chain. Antimetabolites have been detected in sea water, particularly in respect to the production of antibacterial substances by phytoplankton (e.g. Sieburth, 1964). More detailed studies on the release of antibacterial materials by phytoplankton (Duff *et al.*, 1966) have revealed species specific differences and in at least one case (Antia and Bilinski, 1967) this had led to the identification of a lecithinase as being an anti-bacterial agent in the chrysomonad, *Monochrysis lutheri.*

3.2.2 HETEROTROPHIC UPTAKE

The two pathways of heterotrophic uptake shown in Fig. 37 are firstly through the plants (facultative heterotrophs) and secondly through the bacteria and other obligate heterotrophs, such as yeasts and moulds. The most important of these two pathways is undoubtedly the second, and the role of plants as autotrophic (or sometimes auxotrophic organisms) far exceeds their role as heterotrophic organisms.

The ability of marine bacteria to decompose a wide range of naturally occurring organic substrates has been widely demonstrated. These substances include the decomposition of chitin (Hock, 1940; Seki, 1965a and b), cellulose (Waksman *et al.*, 1933), protein (Wood, 1953), and alginates (Meland, 1962). Jannasch (1958) showed that the presence of large numbers of bacteria in sea water was dependent on the concentration of dissolved organic matter; in the absence of soluble substrates, bacteria are found attached to particulate materials, such as inorganic particles and chitin. The heterotrophic activity of bacteria results in a number of complex transformations which are not implied in Fig. 37; these have been reviewed by ZoBell (1962) and include dissolution and precipitation of organic and inorganic compounds as well as the large scale production of inorganic energy reserves (such as methane and hydrogen sulphide), when heterotrophic decomposition occurs under anaerobic conditions (see Fig. 53).

The heterotrophic activity of phytoplankton has been demonstrated in a number of specific cases. Out of 44 pure cultures of littoral diatoms, Lewin and Lewin (1960) found that 28 were capable of heterotrophic growth on glucose, acetate or lactate media. Kuenzler (1965) found that only one out of 13 species of phytoplankton could assimilate glucose for growth in the darkness. Antia *et al.*, (1969) demonstrated that a photosynthetic cryptomonad, *Chroomonas salina*, could grow heterotrophically in the dark at high concentrations of glycerol but not at low concentrations. However, a variety of substrates including acetate, glutamate, and glucose stimulated growth of the organism in the light. Sloan and Strickland (1966) used radioactive carbon to determine the heterotrophic uptake of glucose, acetate, and glutamate by four species of marine phytoplankton. While some uptake was detectable, it was generally less than the dark uptake of radioactive carbon dioxide; the latter can be fixed non-photosynthetically in the dark through biochemical mechanisms, such as the Wood–Werkman reaction (Wood and Werkman, 1936 and 1940). At times the dark uptake of carbon dioxide may exceed the amount taken up by photosynthesis; this indicates that in such environments heterotrophic activity, resulting in the mineralization of organic matter, is in excess of autotrophic activity resulting in the production of organic matter (e.g. Seki, 1968, and Section 3.1.6).

The uptake of organic substrates at low concentrations has been demonstrated among a few species of coastal phytoplankton. North and Stephens (1971 and 1972) showed that *Platymonas* and *Nitzschia ovalis* could take up amino acids. In the case of *Platymonas*, for example, arginine, glycine, and glutamate were taken up at concentrations of *ca* 5×10^{-6} moles/litre; the rate of uptake was increased 10-fold in nitrogen-starved cells. As a general conclusion, however, it appears that heterotrophic activity among the phytoplankton is limited to a number of species and that among these, some species may require very high concentrations of substrate in order to demonstrate heterotrophic activity.

Parsons and Strickland (1962b) introduced the use of radioactive organic substrates to a study of the kinetics of heterotrophic uptake in sea. The author's original experiments showed that the uptake of radioactive glucose and acetate by natural microbial populations in the sea could be fitted to the equation

$$v = \frac{V_m(S+A)}{K+(S+A)} \tag{57}$$

where v was the rate of uptake of the substrate, S was the natural *in situ* concentration of substrate, A was the concentration of the same substrate added to the water, V_m was the maximum rate of uptake and K was a constant, similar to the Michaelis–Menten constant for enzyme reactions (see Appendix to this section). Quantitatively v, in units of $mg/m^3/hr$, was obtained from the expression

$$v = \frac{cf(S+A)}{C\mu t}, \tag{58}$$

where c was the radioactivity of the filtered organisms (cpm), S and A were as defined above, C was the activity of 1 μCi of ^{14}C in the counting assembly used, μ was the number of microcuries added to the sample bottle, f was a factor for any isotope discrimination between the ^{14}C isotope and normal carbon, and t was the incubation time in hours. By employing a reciprocal plot of eqn. (57), to obtain a straight line, the negative intercept on the x-axis becomes $-1/(K+S)$. If one could assume that $K \ll S$, the method would give a measure of the amount of natural substrate in sea water; alternatively, if $K \gg S$ the method would give a measure of the "relative heterotrophic potential" based on K in mg C/m^3, which expresses the ability of the microbial population to take up the substrate. Wright and Hobbie (1965 and 1966) discussed the use of this method as applied to lake water and from their work it has been possible to draw a number of conclusions. By combining eqn. (57) and (58) and rearranging, the following form can be obtained:

$$\frac{C\mu t}{fc} = \frac{(K+S)}{V_m} + \frac{A}{V_m}. \tag{59}$$

When $C\mu t/fc$ is plotted against different concentrations of A (assuming A to be much larger than S), a straight line is obtained with a slope $1/V_m$, an intercept on the ordinate of $(K+S)/V_m$ and an intercept on the abscissa at $-(K+S)$. If an independent determination is made of S (e.g. by chemical analysis), the rate of uptake of the substrate under natural conditions can be determined, providing the organisms are complying with the relationship in eqn. (57). Three experimental qualifications governing the use of this equation are (i) that there must be no appreciable change in the microbial population during the experiment, (ii) that there should be no appreciable change in substrate concentration and, (iii) that careful temperature control is maintained. A plot of $C\mu t/fc$ versus A is shown in Fig. 38

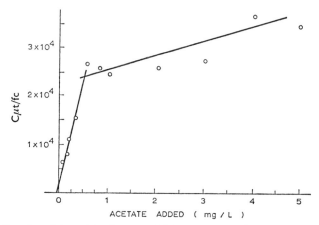

FIG. 38. Kinetics of acetate uptake by natural lake water samples at different substrate concentrations plotted according to eqn. (59) (redrawn from Wright and Hobbie, 1965).

from Wright and Hobbie (1965). From these results with natural lake water it is apparent that eqn. (57) applied over concentration range from 0 to 580 mg/m³ of added acetate, but that above this concentration the relationship did not apply. In additional experiments, it was further shown by Wright and Hobbie (1965) that at high substrate concentrations a second uptake mechanism was involved which accounted for the departure from an asymptotic value of V_m predicated by eqn. (57). This was explained in terms of a passive transport of substrate due to diffusion at high substrate concentrations and the process could be

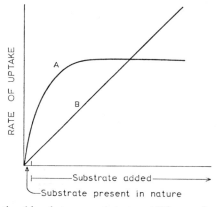

FIG. 39. Theoretical relationships between substrate addition and velocity of heterotrophic uptake showing (A) active transport, in which the transport system becomes saturated and (B) transport by diffusion in which the uptake velocity increases with substrate concentration. Note: Added substrate is greatly in excess of natural levels.

expressed in terms of a diffusion coefficient (K_d) as

$$K_d = \frac{V_1 - V_2}{A_1 - A_2},\qquad(60)$$

where V_1, V_2 and A_1, A_2 are two rates of uptake and substrate concentrations, respectively. The difference in these two uptake mechanisms is illustrated in Fig. 39, curves A and B.

From these studies it is apparent that there are at least two transport systems for heterotrophic uptake: one showing dependence on Michaelis–Menten kinetics at low substrate concentrations and the other showing dependence on diffusion at high concentrations. Some indication of how low substrate concentrations can be in nature for the first mechanism, is given by the K values for two bacterial isolates obtained by Wright and Hobbie; these values were 5·0 mg/m³ for acetate and 7 mg/m³ for glucose. Since 90% of the maximum uptake velocity is obtained at ca $10 \times K$, the active uptake mechanism may be assumed to operate maximally at substrate concentrations below 100 mg/m³. The second mechanism for heterotrophic uptake appears to operate at about ten times the maximum concentration for active uptake (i.e. at ca 1·0 g/m³) and from some additional experiments it was shown by Wright and Hobbie (1965) that phytoplankton were largely responsible for the uptake of organics by the diffusion mechanism. From these findings it was concluded that, in general, bacterial activity in aquatic environments would effectively maintain concentrations of substrates which would be too low for algal heterotrophy, if the latter was dependent only on the diffusion of substrates at high concentrations. However several authors have found that Michaelis–Menten kinetics may apply to the uptake of organic substrates by some species of coastal phytoplankton. For examples, Hellebust (1970) found that one species of marine diatom, *Melosira nummuloides*, showed an ability to take up amino acids through an active transport system which compared favourably (K as low as $7·7 \times 10^{-6}$ M) with transport constants of many bacteria (see also North and Stephens, 1971 and 1972). At present, however, these organisms do not appear to account for heterotrophic activity in oceanic environments. This is indicated from field data collected by Williams (1970), who found that 80% of the heterotrophic activity in sea water samples from the Mediterranean and Atlantic Ocean would pass an 8 µ filter. This would eliminate a large part of the phytoplankton biomass. Particles of <8 µ would include all bacteria and most bacterial aggregates, other than those attached to large particles of detritus.

The hyperbolic curve (A) in Fig. 39 is shown passing through zero rate at zero substrate concentration. In practise it has been found (Jannasch, 1970) that zero growth may occur at some low but finite, or threshold, concentration. This phenomenon has been explained as a population effect in which a certain initial population of a particular species has to be present in order to modify the environment (e.g. redox potential) for the cells to multiply. The effect has been observed in chemostat studies but it may also be a common phenomenon under natural conditions in the sea. Kinetics accounting for a small departure from a zero intercept have been given by Jannasch (1970).

In some cases it has been reported to be difficult to establish any meaningful relationships between substrate concentration and uptake (e.g. Vaccaro and Jannasch, 1967; Hamilton and Preslan, 1970): these difficulties may be due to the existence of several competing heterotrophic populations or to the occurrence of more than one uptake mechanism operative over the same range of concentration. From Hellebust's (1970) data it is also apparent that a form of competitive inhibition of substrate uptake may be caused by the presence of similar substrates.

Sorokin (1970) has investigated the uptake of radioactivate algal hydrolysates in water samples taken at different depths from the surface to 5000 m in the Pacific Ocean. The author found very little heterotrophic activity below 800 m and the maximum activity in tropical waters was between 400 and 600 m, which coincides with the depth of the oxygen minimum. From fairly extensive studies on the uptake of glucose and amino acids in the English Channel, Andrews and Williams (1971) concluded that heterotrophic processes in the area account for an uptake of organic material equivalent to 50% of the measured phytoplankton production. If this is converted into bacterial biomass with an efficiency of between 30 and 60%, then the contribution of heterotrophs to the formation of particulate organic material in the ocean would amount to between 15 and 30% of the production by autotrophic photosynthetic processes.

(Appendix to Section 3.2.2)

In a number of sections throughout this text reference is made to Michaelis–Menten kinetics. Caperon (1967) discussed the general applicability of this approach to the uptake of food materials by microorganisms. From his review of the subject it is possible to write a general expression for the growth of a population of microorganisms limited by food supply. The approach follows the original expression given by Michaelis and Menten for the action of enzymes on substrates.

The process of a population feeding may be described by the following equation in which C is the acquisition site for food, B is a food particle or unit quantity of food, $C \cdot B$ is an acquisition site filled by B, and P is a particle, or unit amount, of ingested food.

$$C + B \underset{k_2}{\overset{k_1}{\rightleftharpoons}} C \cdot B \xrightarrow{k_3} C + P. \tag{61}$$

If $(C+B)$ to $(C \cdot B)$ is a reversible process it can be defined by constant K_m such that

$$K_m = \frac{[(C) - (CB)]\,(B)}{(CB)} \tag{62}$$

or

$$(CB) = \frac{(C)\,(B)}{K_m + B} \tag{}$$

If the breakdown of (CB) to $(C+P)$ is largely an irreversible process, then the rate (v) for this process is given by

$$v = k_3\,(CB) \quad \text{and,} \tag{63}$$

$$v = \frac{k_3(C)\,(B)}{K_m + (B)}. \tag{64}$$

The maximum rate (V_m) will be attained when the concentration of CB is maximized, which is when all the acquisition sites are filled [i.e. $(CB) = (C)$].

$$V_m = k_3\,(CB) = k_3\,(C). \tag{65}$$

By combining eqn. (64) and (65), the general Michaelis–Menten expression is obtained:

$$v = \frac{V_m(B)}{K_m + (B)}. \tag{66}$$

Under certain conditions it is possible to relate growth to the rate of food intake. These conditions are (1) that the concentration of only one food is limiting population growth,

(2) that a given amount of ingested food always results in the production of a fixed number of new individuals, (3) that there is no time lag in the response of growth rate to change in food concentration, and (4) that food once absorbed is ingested and not returned in any appreciable quantity as a food item. Under these conditions, growth, measured as production dn/dt per unit biomass (n) can be expressed as μ, the growth constant, where

$$\mu = \frac{1}{n}\frac{dn}{dt} \tag{67}$$

and μ and μ_{max} substituted for v and V_m in eqn. (66)

$$\mu = \frac{\mu_{max}(B)}{K_m+(B)}. \tag{68}$$

CHAPTER 4

PLANKTON FEEDING AND PRODUCTION

4.1 FEEDING PROCESSES

Among the zooplankton, methods of feeding may be broadly divided into filter feeding and raptorial feeding. In the former process a number of different mechanisms are employed to induce a flow of water against a screen which removes most of the particulate material in suspension. In the second process, animals actually seize individual prey items; the latter may either be large plants, such as individual diatom cells, or planktonic animals generally smaller in size than the predator. The two processes are not mutually exclusive and examples of both filter feeding and raptorial feeding can often be found in one species, especially among the planktonic crustaceans. On the basis of diet, zooplankton feeding may be herbivorous, omnivorous, or carnivorous; other divisions used by some authors are to describe the dietary requirements as being phytophagous, zoophagous, euryphagous, or detritivorous.

A general account of suspension feeding has been written by Jørgensen (1966), and an extensive study of the nutrition of abyssal plankton has been reported by Chindonova (1959); specific reviews on planktonic crustacean feeding have been written by a number of authors including Marshall and Orr (1955), Wickstead (1962) and Gauld (1966).

Among the planktonic protozoans, the feeding of the foraminifera and radiolaria is generally accomplished through the extrusion of pseudopodia from within the main body of the animal; a large number of pseudopodia may spread out into a reticulum and phytoplankton cells are captured on the surface and digested externally (MacKinnon and Hawes, 1961). In addition to capturing prey some of these animals may contain symbiotic algae known collectively as 'zooxanthellae' which may contribute directly to the animal's nutrition through releasing extracellular products of photosynthesis (e.g. Khmeleva, 1967). A second large group of protozoa, the tintinnids, feed off a wide variety of particles including bacteria, detritus, flagellates, and diatoms; food is captured through a large aperture surrounded by cilia which induce a strong current of water towards the animal.

The position of zooplanktonic crustaceans, as by far the largest group of suspension feeders in the ocean, together with their ability to consume a great variety of prey, warrants special consideration of the feeding habits of these animals in the marine food chain. A schematic representation of the location of feeding appendages in a calanoid copepod is given in Fig. 40 adapted from Jørgensen (1966) and Gauld (1966). While the details of these appendages vary considerably among the planktonic crustaceans, their general form and use includes a means of inducing water currents, filtering out organisms, seizing prey, and cutting or grinding food particles; the exact development of these processes is more or less specialized throughout the group. Thus in the illustration shown, a filter chamber is formed by the long setae of the 2. maxillae which project ventroanteriorly towards the labrum covering the mouth. Water currents are produced by a combination of vibrations of the 2. antennae, the mandibles, the 1. maxillae and the maxillipeds. The flow of water is as indicated in Fig. 40

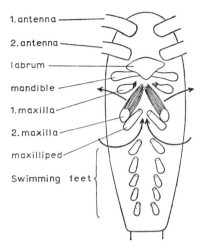

1. antenna
2. antenna
labrum
mandible
1. maxilla
2. maxilla
maxilliped
Swimming feet

FIG. 40. Diagram of ventral view of a copepod showing relative position of major appendages and flow of water (arrows).

and the removal of particular material caught on the 2. maxillae is largely accomplished by a combing action of the 1. maxillae and the maxillipeds which pass the food forward to the mouth. The size range of particles retained by this type of filtering mechanism has been discussed by Gauld (1966) among others. The lower limit appears to be largely determined by the spacing of the setules which are small hair-like structures or bristles on the setae. In *Calanus finmarchicus* the setules of the 2. maxillae are spaced about 5 μ apart. Jørgensen (1966) has summarized reported measurements on a number of different copepod species including various life stages; the range of values for copepods was from 1·5 μ for *Pseudocalanus minutus* to 8·2 μ for *Metridia longa*. Similar structures on euphausiids showed a much greater range from about 7 μ for *Euphausia superba* up to 60 μ for *Bentheuphausia ambluops* (Nemoto, 1967). The largest particle which can be retained by the filter feeding process appears to be decided by the position of the maxillary setae in forming the filter chamber. Gauld (1966) has discussed the details of this structure in *Calanus finmarchicus* and has concluded that particles larger than 50 μ would be excluded from entering. It is quite apparent, however, that many filter-feeding zooplanktonic crustaceans are capable of feeding off diatom cells which are much larger than 50 μ diameter. This is accomplished through a second feeding mechanism in which large diatom cells are grasped by the animal and their contents sucked out (Beklemishev, 1954; Cushing, 1955). The process may be partly the same as that in which a zooplankter captures a smaller animal. However, the capture of large phytoplankters and small zooplankters may differ in respect to the predators feeding response; in *Calanus hyperboreus* Conover (1966c) observed that *C. hyperboreus* could detect the movement of an *Artemia* nauplius which it actively pursued and caught.

The adaptation of crustacean appendages for filter feeding, raptorial feeding or a combination of both, has been discussed by a number of authors. Anraku and Omori (1963) showed that the 2. antennae, mandibles, 1. maxillae and maxillipeds of a predominantly herbivorous filter feeder were all well developed in terms of setae which increased the surface area of these appendages and therefore assured their efficient use in producing water currents through the filter chamber; setae of the 2. maxillae were also well developed in order to assure an efficient particle filter. In contrast, in a predatory species of copepod, *Tortanus discaudatus*, the appendages have few setae and instead there are modifications in structure which aid in the use of these appendages for seizing and holding a prey item. An illustration of two extremes in structural development in *C. finmarchicus* and *T. discaudatus* are shown in Fig. 41. Anraku and Omori (1963) also noted differences in the cutting edges of mandibles

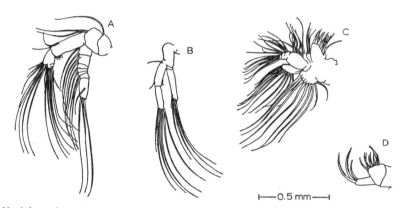

FIG. 41. Adaptation of crustacean appendages showing 2. antenna of (A) *Calanus finmarchicus* and (B) *Tortanus discaudatus* and 1. maxilla (C) and (D) of the same species, respectively (redrawn from Anraku and Omori, 1963).

in different copepods and they describe a typical filter feeder as having grinding teeth while those of a typical predator have very sharp teeth. Between these two extremes, there are a variety of structures which enable some copepods to be omnivorous. Itoh (1970) has been able to summarize the difference between herbivores, omnivores, and carnivores on the basis of an "Edge index" (*EI*) derived from measurements of the cutting edges of the mandible. Calculation of the "Edge Index" is illustrated in Fig. 42A and a plot of this

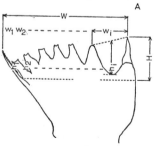

$$\text{EDGE INDEX}: \Sigma\left(\frac{w_i}{W}\cdot\frac{h_i}{H}\cdot 10^4\right)/N$$

FIG. 42A. Schematic representation of the "Edge Index" on the cutting edges of the mandible, $w_1, w_2, \ldots, w_i$ represents the width of individual cutting edges and W the total width; similarly $h_1, h_2, \ldots, h_i$ and H represent the edge heights, respectively; N represents the number of edges (redrawn from Itoh, 1970).

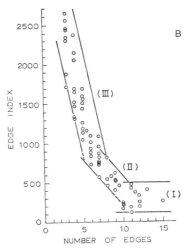

FIG. 42B. Relation between "Edge Index" and number of edges of the mandible (redrawn from Itoh, 1970).

value versus the number of cutting edges, is illustrated in Fig. 42B. From the latter figure it is apparent that pelagic copepods may be divided into three groups viz. Group I, with $EI \leqq 500$ consisting of the Families, Calanidae, Eucalanidae, Paracalanidae, and Pseudo-calanidae, which are largely herbivorous filter feeders; Group II, with $500 < EI \leqq 900$, consisting of generally omnivorous feeders of the Families, Euchaetidae, Centropagidae, Temorida, Lucicutiidae and Acartiidae, and Group III $EI > 900$ consisting of predominantly raptorial feeders of the Families, Heterorhabdidae, Augaptilidae, Candaciidae, Pontellidae, and Tortanidae. From earlier work on stomach contents of these animals it is sometimes possible to identify the raptorial feeders as being carnivores although the inclusion of stomach contents of a herbivorous prey makes it difficult to diagnose whether an animal is exclusively carnivorous.

In addition to extensive studies on the feeding of copepods, Nemoto (1967) has reported on differentiation in the body characters of euphausiids. Among euphausiids, raptorial feeders usually have very long 2nd or 3rd thoracic legs which terminate in grasping spines or small chelae; predominantly filter feeding euphausiids have thoracic legs of similar length and well developed setae. In addition to modifications in the appendages of these animals, Nemoto examined differences in stomach structures and reported that herbivorous euphausiids had a cluster of spines along the posterior wall of the stomach while the same structures were largely absent from deep water carnivorous species.

Among other major groups of zooplanktonic animals, the Coelenterata and Ctenophora are almost exclusively carnivorous and prey are collected largely through passive encounter with the animals' tentacles. However, Jørgensen (1966) has reported that at least one jelly fish is capable of suspension feeding and that the process of prey capture involved the adhesion of particles to the mucus surface of the animal, over which a surface current was induced by ciliary action. Another large carnivorous group of zooplankton are the Chaetognaths, or Arrow worms, which capture their food using specialized jaws. Among the planktonic Heteropods, the pteropod *Limacina* collects food on the flat surface of its foot and within its mantle cavity. The process of food collection is by adhesion of food particles on a mucus membrane which is carried forward into the animal's mouth by ciliary action

(Jørgensen, 1966). Chindonova (1959) has reported that among food items recognized in pteropods, the most common organisms were diatoms, Radiolaria and Globigerinae. The retention of very fine particulate materials is apparent in some members of the Urochordata, which includes the salps and the appendicularians. Among the latter, *Oikopleura* is reported to retain particles down to 0·1 μ in size; this is achieved by ciliary induced currents passing water through a funnel shaped mucus net, the ends of which are twisted into a string which is continuously being consumed through the animal's oesophagus (Jørgensen, 1966).

The rate of filtering among microcrustaceans is broadly related to body size but can vary in any individual depending on such factors as temperature and food concentration. From a summary of measurements made by Marshall and Orr (1955) and Jørgensen (1966) for copepods, it appears that the volume of water filtered by an animal may range from *ca* < 1·0 to 200 ml/day which for a size range of animals from nauplii to adults gives an approximate filtering rate of 40 ml/mg wet weight of animal per day under conditions of active feeding. Cushing and Vucetic (1963) determined that maximum filtering rates for adult copepods could be as high as 1 litre/day. The latter figure appears high, however, when compared to the filtering rates of the much larger euphausiid crustaceans; these have been measured by Raymont and Conover (1961) and Lasker (1966) who report a range of values for *Meganyctiphanes norvegica*, *Thysanoëssa* sp. and *Euphausia pacifica* from *ca* 1 to 25 ml/hr. The filtering rate of marine planktonic crustaceans may decline as food concentration is increased. In experiments conducted by Mullin (1963) adult *Calanus hyperboreus* appeared to filter between 200 and 300 ml/day, decreasing linearly to less than *ca* 50 ml/day as the concentration of food increased. The actual regression equation between ml filtered per

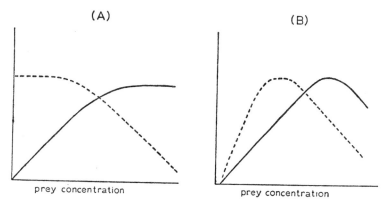

Fig. 43. Two possible effects of the filtering rate (– – –) on the food intake (———) at different prey concentrations. (A) Filtering rate constant or declining with increased prey concentration, (B) filtering rate increasing with prey concentration at low prey densities and decreasing rapidly at high prey densities.

day per copepod and the concentration of phytoplankton depended on the species of phytoplankton and the age of the culture. Although it does not appear in the data given by Mullin (1963), it is also probable that at very low cell concentrations, filtering rates do not continue to increase (Nassogne, 1970), but may even decline. Two possible effects of filtering rate on the intake of food are shown in Fig. 43. In the first illustration a constant maximal filtration rate at low prey concentrations gives an increased food intake with increased prey concentration; the filtering rate then declines at higher prey concentrations to give a maximal food intake. In the second illustration the filtering rate increases with prey concentration, reaches a maximum asymptotic value and declines at higher concentrations. Food intake in the second illustration is zero at some finite concentration of prey, reaches a maximum and then declines.

Among omnivorous microcrustaceans, the filtering rate can be greatly suppressed if the animal changes from filter to raptorial feeding; this has been observed by Lasker (1966) with the filtering rate of *Euphausia pacifica*, which was decreased by an order of magnitude, and in some cases eliminated, when nauplii were offered as prey in the presence of phytoplankton.

Over longer time periods the capacity to feed at any concentration must be governed by the animal's capacity to absorb its food. For example, the chaetognath, *Sagitta hispida*, had a maximum capacity for about 50 *Artemia* nauplii per day regardless of food concentration (Reeve, 1964). Filtering rates generally increase with temperature up to some optimal value for the species and then decline. Recently Kibby (1971) has shown that within a species, the optimum temperature for maximum filtration may depend on the temperature to which the species has become adapted. This is illustrated in Fig. 44.

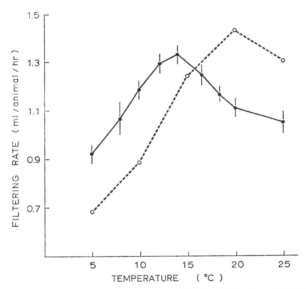

FIG. 44. Filtering rates of *Daphnia rosea* grown at 12°C (solid line) and 20°C (broken line)— (redrawn from Kibby, 1971).

The filtering rate (F) of a zooplankter over a given time period (t) can be measured as the decrease in phytoplankton cell concentration at the beginning (C_0) and end (C_t) of the experiment (Gauld, 1951). If the assumption is made that the concentration of cells decreases exponentially with time, then

$$F = \frac{v(\log C_0 - \log C_t)2 \cdot 303}{t},\tag{69}$$

where v is the volume of water per animal. A possible error in the use of this expression is that it assumes that F is constant over the period of the experiment. However, since the filtering rate is known to vary with the concentration of cells, the assumption that F is constant is only an approximation which will be true for small changes in $C_0 - C_t$. Control flasks, in which changes in C_0 can be measured, and gently stirred or rotated incubation chambers which assure an even distribution of cells, are necessary refinements in most experiments. A discussion of the methodology of measuring filtering rates has been given by Rigler (1971).

In conclusion it appears from what is known of feeding processes among the zooplankton that the smallest particles ($ca\ 0.1\ \mu$) are retained by animals possessing a mucus net through which water is pumped by ciliary action. Larger particles ($ca > 5\ \mu$) are retained by filter feeding copepods and some other crustaceans while very large particles ($ca > 50\ \mu$) are retained by raptorial feeding; the latter process includes the capture of many of the smaller zooplankton which serve as food for larger members of the plankton community.

4.2 SOME TROPHODYNAMIC RELATIONSHIPS AFFECTING THE PLANKTON COMMUNITY

4.2.1 FOOD REQUIREMENTS OF ZOOPLANKTON

The food requirements of an animal can be expressed in terms of R, the food consumed, G, the food used for growth (and reproduction), T, the food used for metabolism within the body and E, the food taken in but not assimilated (Richman, 1958).

$$R = G + T + E. \tag{70}$$

Excluded from this equation is a small metabolic loss of food which can be identified as the urinary loss in higher animals but which would be included as a small fraction of E in the above equation for zooplankton.

Since E can be expressed as $(R - AR)$ where A is the assimilation efficiency of the food, eqn. (70) can be written as

$$G = AR - T. \tag{71}$$

The extent to which food is assimilated by a zooplankter is difficult to measure experimentally since there are technical difficulties in making quantitative collections of faeces from zooplankton, even under laboratory conditions. By assuming that the ash content of the food material was largely unabsorbed in the digestive process, Conover (1966a) derived a formula which allows for an estimation of the assimilation efficiency (A) in terms of (J), the ash-free dry weight to dry weight ratio in the ingested food and (L), the same ratio in a sample of faeces.

$$A = \frac{(J-L)}{(1-L)(J)} \times 100\ (\%). \tag{72}$$

Use of this formula avoids the difficulty of quantitatively collecting faeces and also permits measurement of assimilation efficiencies on samples of particulate material collected *in situ*. Conover (1966b) examined factors which affect the assimilation efficiency and found it to be largely independent of temperature, the amount of food offered, or the amount of food consumed. The assimilation efficiency was affected by the ash content of the phytoplankton food, however, and a simpler empirical relationship than the one described above was established, such that

$$A = 87.8 - 0.73X, \tag{73}$$

where (X) is the percentage ash per unit dry weight of food. From a number of experimental estimates of (A), as well as from additional reports in the literature, Conover (1968) concluded that in general zooplankton assimilation efficiencies lie in the range 60 to 95%.

The determination of assimilation efficiencies based on the ash content of food and faeces may be a useful method for determining organic carbon assimilation but it appears that the assimilation efficiency of elements, such as nitrogen and phosphorus, cannot be determined with any accuracy by this method (Kaczynski, personal communication). The quantities of these elements absorbed are much smaller than the quantities of organic carbon and small changes in the ash content of the food and faeces can greatly effect the calculated elemental assimilation efficiency. An alternative method which relies on the ratio of elements to each other (e.g. N to P) in the food and faeces is given by Butler *et al.* (1970) and Kaczynski (in press).

Food used for metabolism within an animal (T) can be expressed as a function of the animal's body weight (W) such that

$$T = \alpha W^\gamma$$

or

$$\log T = \log \alpha + \gamma \log W. \tag{74}$$

T can be measured in terms of an animal's respiration (e.g. $\mu\,1\,O_2$/animal/hr) and the value of α will be a function of the units which are used to express the animal's weight (e.g. wet weight, dry weight etc.) as well as being influenced by environmental factors such as temperature. γ, on the other hand, has been found to be relatively constant (Zeuthen, 1970) and to reflect the internal metabolism of the organism; since there are certain overall similarities in cellular metabolism of a large variety of species, the relative constancy of γ appears logical. However, from a very extensive survey of different zooplanktonic organisms, including crustaceans, pteropods, chaetognaths, coelenterates and polychaetes, Ikeda (1970) established that there was a significant difference between the value of γ for tropical, temperate, and boreal species taken from the tropical Pacific, the temperate Pacific southeast of Hokkaido, and the Bering Sea, respectively. The results of Ikeda's studies are shown in Table 24 and Fig. 45. These results, in which metabolism is expressed per unit weight of animal, clearly show that the smallest animals have the highest metabolic rate and that there are geographic differences, not only in the value of α, but also in the value of γ, which reflect an environmental difference in basic metabolic processes of animals from different areas.

TABLE 24. REGRESSION EQUATIONS OF LOG SPECIFIC RESPIRATION RATE, R' (μlO_2/mg BODY wt/hr) AND LOG BODY WEIGHT EXPRESSED AS THE DRY WEIGHT OF THE ANIMAL (mg/ANIMAL) DERIVED FROM DATA IN FIG. 45 (IKEDA, 1970).

Species	Regression equation
Boreal	$R' = -0.169W + 0.023$
Temperate	$R' = -0.309W + 0.357$
Tropical	$R' = -0.464W + 0.874$

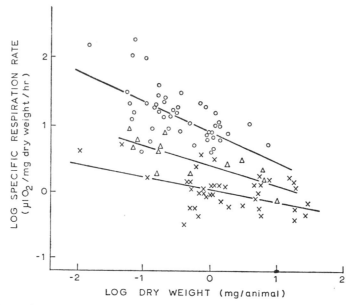

FIG. 45. Regression lines for planktonic animals from boreal ($\times$), temperate ($\triangle$) and tropical ($\bigcirc$) waters drawn for log. specific respiration rate (μl O$_2$/mg dry wt./hr.) against log. dry weight animal (mg)—(redrawn from Ikeda, 1970).

For the same species of animal at different temperatures (i.e. within species variation), Comita (1968) determined a multiple regression equation for *Diaptomus* which gave the oxygen uptake (T in μl/mg dry wt/hr) in terms of the temperature (t in °C) and the dry weight (W in mg) as follows

$$T = 0 \cdot 0364t - 0 \cdot 3418 \log W + 0 \cdot 6182. \tag{75}$$

The conversion of metabolism, which is generally measured as respiration in units of oxygen consumed, to biomass of food in terms of organic carbon is made by the following equations

$$\text{mg O}_2 \text{ consumed per unit time} \times \frac{12}{32} \times RQ$$

$$= \text{mg C utilized per unit time; \quad or}$$

$$\text{ml O}_2 \text{ (NTP) consumed per unit time} \times \frac{12}{22 \cdot 4} \times RQ$$

$$= \text{mg C utilized per unit time}$$

where RQ is the respiratory quotient, $+\Delta CO_2/-\Delta O_2$, which may range in animals from $0 \cdot 7$ to $1 \cdot 0$ depending on whether fats or carbohydrates are being utilized for energy. Thus in a copepod living off fat reserves, an RQ of $0 \cdot 7$ may be recommended, while a copepod feeding off phytoplankton may have an RQ closer to $1 \cdot 0$.

The food consumed by a zooplankter (R) can be expressed in terms of the amount of food available. Using a variety of fish, Ivlev (1945) determined that the quantity of food eaten increased with the concentration of food offered, up to some maximum ration which could not be further increased by increasing the concentration of food. From this observation it was stated that if the maximum ration is taken as R_{max}, then the relation between the size of the actual ration, R, and the concentration of prey, p, must be proportional to the difference between the actual and maximal ration such that

$$\frac{dR}{dp} = k(R_{max} - R) \tag{76}$$

where k represents a proportionality constant. Integrating, this expression gives

$$R = R_{max}(1 - e^{-kp}). \tag{77}$$

Sufficient evidence exists in the experimental data from a number of authors (e.g. Reeve, 1963; Mullin, 1963; Parsons et al., 1967; Sushchenya, 1970, and references cited therein) to show that the general form of this relationship can be applied to zooplankton grazing. Two modifications to this expression have been proposed on the basis of additional experimental results. The first of these was shown by Ivlev (1961) to apply to the feeding of fish under conditions in which the effect of prey aggregation on feeding was to increase the ration obtained when the prey concentration was held constant. It is difficult to establish on the microscale of zooplankton whether this effect with fish is applicable to zooplankton feeding. It is apparent, however, that natural mechanisms which tend to concentrate prey cause patchy distributions of both plants and animals (see Section 1.3.5), and that from the observation given above, the occurrence of patchy distributions may increase the food available when the average prey concentration is quite low (Paffenhöfer, 1970). The second modification to eqn. (77) concerns the prey concentration at which feeding starts. From observations on zooplankton grazing conducted at sea (Adams and Steele, 1966; Parsons et al., 1967), as well as from laboratory experiments (e.g. Nassogne, 1970), it appears that grazing ceases at some minimum or threshold prey concentration, p_0. Equation (77) can be modified to include this value as

$$R = R_{max}(1 - e^{k(p_0 - p)}). \tag{78}$$

Threshold values for p_0 have been reported to range from 40 to 130 µg C/l depending on species of microcrustacea and food (Parsons and LeBrasseur, 1970). In other experiments, however, it appears that filtering continues down to zero prey concentration (e.g. Mullin, 1963; Paffenhöfer, 1970). In the latter report it is shown that the copepod, Calanus helgolandicus, could attain its maximum ration (R_{max}), as determined from the largest sized animals, at food concentrations of between 100 and 200 µg C/l.

The actual quantity of food eaten by zooplankter (R) depends on the type of prey as well as its concentration; thus E. pacifica obtained 15% of its body weight per day off a bloom of Chaetoceros but at equivalent concentrations of nanoplankton, E. pacifica could only obtain approximately one third of this amount of food per unit time (Parsons et al., 1967). In this example it may be assumed that the small nanoplankters (8 µ diameter) were much less efficiently filtered than the larger diatoms (32 µ diameter). Prey selectivity may also determine the amount of food eaten. Conover (1966c), and others have observed for example that some food particles will be rejected by zooplankton even after they have been seized during raptorial feeding. Also it has been shown that within the same species of phytoplankton, larger food particles, consisting of paired cells, may be selectively grazed over smaller single cells (Richman and Rogers, 1969). The degree to which an animal selects a prey over the natural abundance of different prey in the environment can be expressed as an electivity index, such as that given by Ivlev (1961) where

$$E = \frac{r_i - p_i}{r_i + p_i} \tag{79}$$

and r_i is the relative proportion (e.g. percent) of a prey in the ration and p_i is the relative proportion of the same prey in the water. Prey concentration can be expressed as units of biomass or numbers of organisms and the choice of units will affect the value of E. The absolute value of E in the above expression ranges from -1 to $+1$, higher values denoting a greater selection of a prey item over lower values.

In general it appears that food ingested (as indicated by R in experiments) has been found to range from a few percent up to *ca* 100% of the body weight of zooplankton per day, with lower average values of 10 to 20% of the body weight per day for the larger crustacean zooplankton and 40 to 60% of the body weight per day for the smaller crustacean zooplankton (Bell and Ward, 1970; Sushchenya, 1970; Mullin, 1963; Parsons and LeBrasseur, 1970). From these references it is also apparent that prey densities at which zooplankton reach the asymptotic value of R_{max} appear to be in the range 200 to 1000 μg C/l, depending (as in the case of p_0) on the particular combination of predator and prey. Inhibition of zooplankton food intake at high prey concentrations appears in some experimental data. From Mullin's (1963) results it was shown that the number of diatom cells (*Thalassiosira fluviatilis*) ingested by *C. hyperboreus* increased over the range *ca* 200 to 4000 cells/ml but then decreased to about 25% of their maximum intake over the range, 4000 to 8000 cells/ml.

Apart from the obvious need for major metabolites such as proteins, carbohydrates and fats, there are other properties of diets, as well as properties of the sea water in which animals reside, which may influence an animal's growth and survival. These properties are at present poorly defined but they are apparent from a number of examples. Thus Wilson (1951) showed that *Echinus* larvae survived better in some waters than others; Provasoli *et al.* (1959) showed that only one or two algae were satisfactory as food for *Artemia* and *Tigriopus*, while a combination of several algae generally proved a much more adequate diet. Lewis (1967) showed that a mixture of trace metals improved the survival of *Euchaeta japonica* and that periodicity in the occurrence of natural chelators in seawater was a possible factor affecting *E. japonica* survival, *in situ* (Lewis *et al.* 1971).

In eqn. (71) the food available for growth (G) is achieved when the gain in food (AR) exceeds the loss (T). This indicates that for a constant basal metabolism (T) a linear increase in growth can be achieved by increasing the ingested ration (AR). However in a review of published data on the feeding of fishes, Paloheimo and Dickie (1965, 1966a and b) found that increasing the ration (R) resulted in a decreased efficiency of food utilization for growth and that the relationship between gross growth efficiency (K_1) and ration was log-linear, such that

$$\ln K_1 = -a - bR \tag{80}$$

where a and b are constants, characteristic of the predator and its prey, and gross (K_1) or net (K_2) growth efficiency are defined as

$$K_1 = \frac{\Delta W}{R \, \Delta t} \quad \text{and} \quad K_2 = \frac{\Delta W}{AR \, \Delta t}.$$

Since $\Delta W / \Delta t$ represents growth per unit time it follows that

$$\frac{\Delta W}{R \, \Delta t} = e^{-a - bR} \tag{81}$$

or
$$\frac{\Delta W}{\Delta t} = Re^{-a-bR} \tag{82}$$

which is an expression of growth (G) in terms of ration (R). Substituting this into eqn. (71),

$$Re^{-a-bR} = AR - T$$

or
$$T = R(A - e^{-a-bR}). \tag{83}$$

The application of Paloheimo–Dickie equations to zooplankton feeding has been discussed by Conover (1968) and Sushchenya (1970) who have concluded from experimental data on zooplankton that the approach may have general application to the study of food requirements of zooplankton. Thus Sushchenya (1970) showed from Richman's data that there was a linear relationship between the logarithm of the gross growth efficiency and ration in the feeding of *Daphnia pulex* and that the maximum gross growth efficiency was 60%. Further it was possible to derive an equation [eqn. (83)] for *Daphnia* feeding on *Chlamydomonas* which closely fitted the experimental growth data. Sushchenya (1970) accumulated various values for the gross and net growth efficiency, and the food used for respiration (T) as a percentage of the food assimilated (AR); the values are reproduced in Table 25. From

TABLE 25. VALUES OF ENERGY COEFFICIENTS K_1 AND K_2, AND PERCENTAGE OF ENERGY EXPENDITURE FOR RESPIRATION (T), IN % OF AR (FROM SUSHCHENYA, 1970).

Species	t °C	K_1	K_2	T	Author
Artemia salina	25	18·5	23·6	76·4	Sushchenya, 1962
Artemia salina	25	13·0	26·5	73·5	Sushchenya, 1962
Artemia salina	25	9·0	27·4	72·6	Sushchenya, 1962
Daphnia pulex	20	13·2	55·4	44·6	Richman, 1958
Daphnia pulex	20	9·1	57·6	42·4	Richman, 1958
Daphnia pulex	20	4·8	56·9	43·1	Richman, 1958
Daphnia pulex	20	3·9	58·7	41·3	Richman, 1958
Calanus helgolandicus	10	48·9	52·5	47·5	Corner, 1961
Calanus helgolandicus	10	37·2	47·0	53·0	Corner, 1961
Calanus helgolandicus	10	41·9	46·2	53·8	Corner, 1961
Asellus aquaticus	14–19	28·3	40·4	59·6	Levanidov, 1949
Asellus aquaticus	14–19	23·3	33·3	63·7	Levanidov, 1949
Asellus aquaticus	14–19	20·6	30·0	70·0	Levanidov, 1949
Asellus aquaticus	14–19	18·7	26·7	73·3	Levanidov, 1949
Asellus aquaticus	14–19	16·3	23·3	76·7	Levanidov, 1949
Euphausia pacifica	10	7·1	7·4	92·6	Lasker, 1960
Euphausia pacifica	10	31·8	39·2	60·8	Lasker, 1960
Euphausia pacifica	10	14·2	14·9	85·1	Lasker, 1960
Euphausia pacifica	10	28·0	29·0	70·0	Lasker, 1960

these results it is apparent that gross efficiencies vary over a wide range but the K_1 for most animals lies in the range of *ca* 10 to 40%; K_2 values are less variable and show a range of *ca* 25 to 55%, while the expenditure of food on metabolism generally lies in the range 40 to 85% of the food assimilated.

The significance of the values (a) and (b) in eqn. (80) has been discussed by Paloheimo and Dickie (1966b). Neither value was affected by temperature which is known to affect the level of metabolism. This was interpreted as meaning that temperature affected the total rate of turnover of food but not the distribution of food among the various metabolic components. By contrast, changes in factors such as salinity and type of food (especially particle size) influenced both parameters.

Several food budgets involving the terms G, AR and T in eqn. (71) have been derived for microcrustaceans. Some results given in Table 26 from Petipa (1966) show that the

TABLE 26. DISTRIBUTION OF FOOD EATEN AS A % OF THE BODY WEIGHT IN *Acartia clausii* AT 17–25°C ASSUMING AN ASSIMILATION (A) OF 80%. (FROM PETIPA, 1966).

	Dry weight (mg)	Daily growth (%)	Daily metabolism (%)	Daily ration (%)
Nauplii	0·000 09	20·3	98·3	148
Stage V	0·002 56	8·3	49·8	73

TABLE 27. DISTRIBUTION OF FOOD REQUIREMENTS IN A EUPHAUSIID FROM LASKER (1966).

Animal	Dry wt. (mg)	Growth Final wt. / Initial wt.	Time (days)	% food assimilated (A)	% food for Growth (G)	% food for moults (G')	% food for eggs (G'')	% food for metabolism (T)	Location of Study
E. pacifica	1·19	3·95	69	86	30	8·0	0	62	Laboratory
E. pacifica	1·65	1·16	63	78	6·2	7·0	0	86·8	Laboratory
E. pacifica	0·23	39·2	580	—	9·4	15·3	8·9	66·4	Natural population

growth and rate of metabolism of *Acartia* nauplii are two or three times greater than for stage V copepods, and that consequently the food intake per unit body weight of the young animals is considerably greater than for the older animals. These results are similar to results obtained for euphausiids (Table 27) which show that faster growing animals devoted proportionally more energy to growth than metabolism, when compared to slower growing animals. In the case of euphausiids, however, an appreciable fraction of up to 15% of the food is devoted to moults. This latter figure may be higher in euphausiids than in copepods. Corner *et al.* (1967) estimated < 1% of the food assimilated by copepods was lost to moults; Mullin and Brooks (1967) showed that *Rhincalanus* lost *ca* 6% of its body carbon as moults during its life cycle from egg to adult. Egg production (G'') may be considered as a second subfraction of 'growth' and in the case of a natural population of *E. pacifica*, this is seen to amount to about 10%.

Throughout the above discussion reference has been made to 'food' requirements and its metabolic distribution. Food may be better expressed in terms of energy units, and the following conversion factors can be used as general approximations:

1 ml of O_2 consumed ≡ 5 g cal assuming,
RQ of approximately 1·0,
1 mg of food (dry wt.) ≡ 5·5 g cal.

assuming an average dry weight composition of food as 50% protein, 20% fat, 20% carbo-hydrate, 10% ash, and caloric equivalents of 5·65 cal/mg for protein, 4·1 cal/mg for carbo-hydrate and 9·45 cal/mg for fat.

The effect of temperature on the metabolic rate (T) has been discussed earlier in this section. However, since temperature affects such overall processes as growth, reproduction, and the final size of organisms, it is appropriate to consider whether there is a general relationship between such complex biological processes and temperature.

McLaren (1963, 1965, and 1966) has discussed the general acceptability of empirical relationships between temperature and the rate of biological reactions with particular reference to the zooplankton community. From his discussions it appears that the two extremes in such approximations are firstly that biological reactions increase in rate two to threefold with a 10 °C rise in temperature; this is described by McLaren as a clumsy approximation which may be assumed to have been adopted for its mathematical simplicity. At the other extreme, mathematically complex polynomial equations can be devised to fit

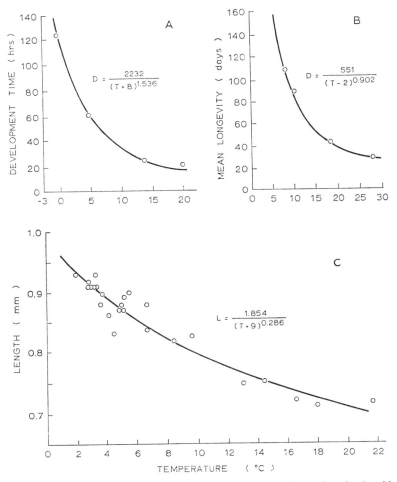

FIG. 46. Effect of temperature on biological development. (A) Time taken for hatching eggs of *Calanus finmarchicus*, (B) Mean longevity of *Daphnia magna*, (C) Mean cephalothorax length in female *Acartia clausi* (redrawn from McLaren, 1963).

almost any deviations in different groups of data. However, from his own data, as well as from a wide variety of published reports, McLaren concluded that the closest fit was generally obtained with Belehradek's empirical formula which is given as

$$V = [a(t + \propto)^b \qquad (84)$$

where V is the rate of a metabolic function, t is the temperature and a, b and $\propto$ are constants. When other factors could be assumed not to be limiting, (e.g. food supply) this function gave good descriptions of development rate, metabolic rate and size (McLaren, 1963); a number of examples are shown in Fig. 46 A, B, and C (note that metabolic rate, V, has been expressed in these figures in terms of time taken to complete a process or length achieved as adults i.e. as $1/V$ on the ordinate). Some physiological and environmental justifications for the use of this function have been discussed by McLaren in terms of the 'constants', a, b and $\propto$. Thus the scale correction, $\propto$, is known as the "biological zero" and expresses the temperature at which $V = 0$. The value $\propto$ appears to be positively related to environmental temperature and consequently has been shown to vary with latitude (and altitude in the case of terrestrial lake communities); the lowest values of $\propto$ being found in the most northerly populations. The constant, b, reflects the degree of curvature and represents the general dependence on temperature of all metabolic processes leading up to changes in the measured parameter, V. Thus when the same metabolic function (e.g. egg hatching) is measured among similar groups of organisms, the curvilinear response, b, should be the same for all groups. The proportionality coefficient, a, is determined by the units in which V is measured; however it is also apparent the value of a can be related to size in such a way as to indicate that it is an expression of a surface to volume restriction governing the exchange of gases across a membrane. This was illustrated by McLaren (1966) who showed a positive correlation between a and egg diameter for a variety of copepod eggs, corrected for differences in their yolk content.

4.2.2 THE MEASUREMENT OF PRODUCTION

Production is defined as the total elaboration of new body substance in a stock during a unit time, irrespective of whether or not it survives to the end of that time (Ricker, 1958). Production depends on the time interval over which it is measured, the presence or absence of predators, and the growth and natural death rate of the population. The combination of these factors is very difficult to measure *in situ* and consequently production estimates are only approximate.* The simplest relationship for the estimation of production is to consider the *average* standing stock of the population and multiply this by an estimate of the generation rate, both determined over short time intervals. For example if the average standing stock of zooplankton is 1 g/m^2 and the average generation rate (or doubling time) is 1 month, then the annual production is 12 $g/m^2/yr$. The obvious difficulty in making this determination is in obtaining figures for the average standing stock and generation rate of a zooplankton population *in situ*.

A direct experimental measurement of potential production can be made in the case of primary producers because their growth is generally rapid so that a change in the population over a short interval (e.g. several hours) can be made in an isolated sample using a variety of sensitive techniques, such as the uptake of radioactive $^{14}CO_2$. In the case of animals,

* *Note*: Primary productivity estimates as measured by the $^{14}C-$ or oxygen techniques actually represents the *potential* production in the absence of sinking, grazing etc.

however, direct measurement of production is difficult although *growth rate* of fish, or zooplankton can be measured over extended periods of time in the laboratory and the information applied to field conditions.

Where a discrete population (or cohort) of animals can be enumerated and the average weight of an individual determined, then the production (P_t) of the cohort over a short time period (t) is given by Mann (1969):

$$P_t = (N - N_t) \times \frac{(\overline{W} + \overline{W}_t)}{2} \tag{85}$$

where N and N_t are the number of animals alive at the beginning and at time t, and $\overline{W}$ and $\overline{W}_t$ are the respective mean weights of the animals.

In this expression the choice of a *short* time period over which to measure changes in W and N really depends on the growth and mortality rates of the population since over long periods growth tends to be exponential. As a very approximate guide for animal populations, a short time interval over which to measure a growth stanza should be ten percent or less of the total generation time of the animal. For a zooplankton population maturing from egg to adult in 3 months, growth observations using eqn. (85) could be made over *short* time intervals of about a week to 10 days.

Equation (85) represents the production of the surviving population; the production removed by predators (and natural mortality) will be the number of individuals lost per unit time, multiplied by the average weight of those lost. If the production removed by predators is determined over short time intervals and the losses totalled for the year, then the total production can be determined as the sum of the surviving population at the end of the year plus the sum of the total lost to predators.

For a population of copepods having one generation per year the growth rate can be determined *in situ* from changes in the weight of two different stages and the time interval between the maximum numbers of each stage (Cushing, 1964; Parsons *et al.*, 1969). This is illustrated in Fig. 47 where the time interval between the maximum number of Stage I and Stage III copepodites is given as 44 days; the mean weight of the respective stages was 0·15 and 0·60 mg. Since growth over an extended period tends to be a logarithmic function

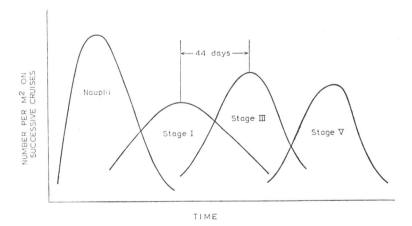

FIG. 47. Estimation of the time intervals in the maximum occurance of different stages of a copepod having one generation (schematic representation of data from Parsons *et al.*, 1969).

of time, the growth rate per day can be approximated from the expression

$$\% \; \varDelta W = 100[10^{1/t(\log W_2 - \log W_1)} - 1] \tag{86}$$

where $\% \; \varDelta W$ is the percent increase in weight per day, W_1 and W_2 are the weights of Stage I and III copepodites respectively and t is the time interval. For the values given above, $\% \; \varDelta W = 3 \cdot 5$ (Parsons *et al.* 1969). From this growth rate the potential production of Stage I copepodites can be determined using $N_1 \times W_1$ (as the total biomass of the cohort) and determining $N_2 \times W_2$ from eqn. (86). The difference between the mean *in situ* biomass of copepods $(\overline{N_2 W_2})$ and the calculated value of $N_2 W_2$, will represent an estimate of the production lost to predators and natural mortality.

In a population of zooplankton which has more than one generation per year it is more difficult to determine production since several generations of the species may be present and the time of development for a single stage can not be determined as in Fig. 47. However if the development time of a stage (or the individual daily growth) is found experimentally (e.g. by isolating different stages and incubating them under *in situ* conditions of temperature and food supply), then the production of the population can be calculated as the sum of the individual stages. In this technique it may not be necessary to deal with the species in stages, and weight groups are in fact more satisfactory for the purposes of the calculation. An example of this determination was made by Grenze and Baldina (1964) and is reported in Table 28 from Mann (1969).

TABLE 28. PRODUCTION OF *Acartia clausii* DURING THE
SUMMER (FROM MANN, 1969).

Size Groups (mg $\times 10^{-3}$)	Number per m³	Individual daily growth (mg $\times 10^{-3}$)	Daily production (mg $\times 10^{-3}$)
0–2·5	850	0·09	76·5
2·5–10	250	0·45	108·0
10–20	87	1·20	104·0
20–30	35	1·68	58·8
30–50	47	1·20	56·4
50–70	40	0·47	18·8
		Total	422·5

4.2.3 THE PELAGIC FOOD WEB AND FACTORS AFFECTING ITS PRODUCTION AND STABILITY

A diagrammatic illustration of a pelagic food web involving the marine plankton community is shown in Fig. 48. Individual species of plankton and fish are illustrated by number up to the i-th representative of each group in any environment. Feeding relationships between predators and prey are illustrated by lines between different trophic levels. Since organisms tend to feed on a diversity of prey, the arrangement of lines is drawn quite arbitrarily to represent interrelationships between organisms. Thus F_1 may be a planktivorous fish which feeds both on herbivorous and carnivorous zooplankton while F_2 may

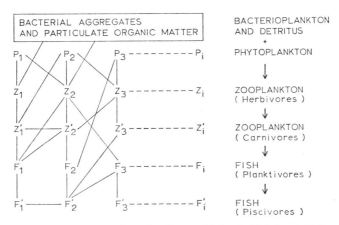

FIG. 48. Diagram of a pelagic food web involving five trophic levels (right hand side) and illustrating connections between different species of organisms, P_1, P_2, P_3, ..., Z_1, Z_2, Z_3, ... etc. at each trophic level.

represent a fish, such as the anchovy, which is capable of feeding on both zooplankton and phytoplankton. The position of the bacterioplankton and detritus is illustrated as providing a source of food for some herbivorous zooplankton but the definition of herbivore, detritivore or carnivore is not exact and some species of zooplankton may take advantage of all three types of prey.

While Fig. 48 illustrates a hypothetical food web, in which the different trophic levels influence each other, there are also a large number of environmental factors which influence both the occurrence and productivity of organisms. These factors such as light, temperature, mixing, and nutrients have been discussed separately in the previous sections and ideally they should be included, together with all the trophic relationships, in any description or model of a marine food web. However to include all the components of variability in such a description is an impossible task. Thus a description of the food web in any environment has to be resolved into its major components. One approach to this problem is to reduce the complexity of a food web into several trophic levels; these are illustrated on the right-hand side of Fig. 48. This simplification is sometimes referred to as a food chain since by analogy the predators and prey are connected by a succession of links up to the highest trophic level. The first step in the food chain is generally thought of as being represented by phytoplankton production. From a trophic position, however, it should also include detrital material, together with the associated bacterioplankton, since these materials form part of the food of herbivorous and omnivorous zooplankton at the secondary level of production. From the secondary level of production food may be transferred to higher trophic levels, depending on the length of the food chain in any environment. One important factor in determining how much food is transferred up to the next trophic level is the 'ecological efficiency' (E); this is defined as

$$E = \frac{\text{Amount of energy extracted from a trophic level}}{\text{Amount of energy supplied to a trophic level}} . \tag{87}$$

Slobodkin (1961) indicated that this value was about 10%. However it is very important to know the correctness of this value since it can greatly influence final estimates of production in which man is interested. For example Schaefer (1965) considered the effect of ecological efficiencies ranging from 10 to 20% on the production of fish at the fifth trophic level. The general form of this production (P) can be given as

$$P = BE^n \qquad (88)$$

where B is the annual biomass at the primary trophic level, E is the ecological efficiency and n is the number of trophic levels. Starting with an annual production of phytoplankton of 1.9×10^{10} metric tons of carbon, the production of fish at the 5th trophic level (3rd carnivore; $n = 4$) was 1.9×10^6 and 30.4×10^6 metric tons of carbon at $E = 10\%$ and $E = 20\%$, respectively. Thus a doubling of the ecological efficiency can result in an order of magnitude increase in the production of a terminal resource.

However, the ecological efficiency as defined above does not take into account recycling processes which take place in nature, but which would not be apparent in a single predator/prey relationship. Thus a 10% transfer of material between trophic levels actually includes a 90% loss to the system. Part of this loss will be real, in that organic carbon will be respired back to CO_2 at each trophic level. However another fraction of the 90% loss will appear as organic debris and be recycled through the food chain (e.g. as indicated in Fig. 37 and discussed in Section 3.2.1). Further it is apparent that at least at the herbivore level, experimental data has recently been presented to show that the gross growth efficiency of copepods (K_1 see Section 4.2.1) from nauplius to adult may be as high as 30 to 45% (e.g. Mullin and Brooks, 1970). Ecological transfer efficiencies will not approach these values for various reasons connected with life cycles, the variety of organisms included in natural trophic levels, predator/prey electivity and the need for wild organisms to spend more energy in hunting their prey. Nevertheless it is apparent that the value of 10% is probably too low for all trophic levels and that it is especially low as an ecological efficiency involving transfers from the first trophic level. For overall estimates it may be assumed that the ecological transfer efficiency at the herbivore level is probably no less than 20% and that transfer efficiencies at higher trophic levels are probably between 10 and 15%. From eqn. (88) it is also apparent that the number of trophic levels (n) occurs as an exponent and as such it can greatly affect production. In an examination of differences in fish production throughout the world, Ryther (1969) considered the number of trophic levels in three communities which may be described as "oceanic", "continental shelf", and "upwelled". Over large areas the author suggested that oceanic communities have long food chains with low transfer efficiencies, the latter being determined by the three or four levels of carnivorous feeding. As an example of such a community the author gave the following food chain which is summarized here as a diagram.

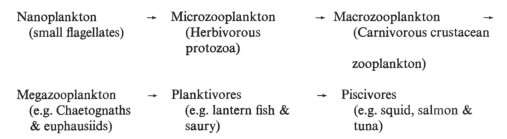

The food chain is shown as a continuous flow of biomass from phytoplankton to fish but in fact salmon, for example, may also feed on euphausiids and squid; thus their position at the 6th trophic level would actually be better represented as being between the 5th and

6th trophic levels. For purposes of comparison with other environments, however, it is apparent that there are approximately five trophic levels leading to the production of a commercial species of fish. Further it was determined from the literature that the annual primary production of oceanic areas was generally low and 50 g C/m²/yr was considered to be an average value.

The second food chain given by Ryther was described as "coastal" but should be more properly called "continental shelf" since it may sometimes exist at considerable distance from the coast, such as in the Atlantic on the Grand Banks. This food chain may be represented diagrammatically as follows:

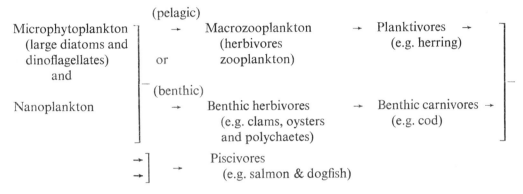

As in the oceanic food chains, the flow of food represented above is an average picture and may include such steps as carnivorous zooplankton in addition to those shown. Essentially, however, the food chain can be represented by three trophic levels, whether this is via the benthic or pelagic community. The total annual primary production of these areas was assessed as *ca* 100 g C/m²/yr.

Finally the food chain of areas in which there is persistent upwelling, such as off the coast of Peru or in Antarctic seas, could be represented by the following diagram:

Macrophytoplankton
(large diatoms and dinoflagellates
including chain forming species)
$-\bigg[\begin{array}{l}\rightarrow\ \text{Planktivores (e.g. anchovy)} \\[4pt] \text{or} \\[4pt] \rightarrow\ \text{Megazooplankton} \rightarrow \text{Planktivores} \\ \quad\ \text{(e.g. } \textit{Euphausia} \qquad \text{(e.g. whales)} \\ \quad\ \textit{superba}) \end{array}$

In this food chain it is known that certain fish, such as adult anchovy, may feed directly off phytoplankton, while another relatively short food chain may exist between euphausiids and whales. The total number of trophic levels in such environments was represented by Ryther as one and a half. The total primary productivity of this community type was assessed at 300 g C/m²/yr since the process of upwelling leads to a continual supply of nutrients.

In these three communities ecological efficiencies at each trophic level were assumed to be highest, when governed largely by phytoplankton/herbivore associations and lowest for communities in which there were secondary and tertiary carnivores. Consequently Ryther assigned a 10% overall efficiency to the oceanic food chain, a 15% efficiency to the continental shelf food chain, and a 20% efficiency to the food chain in upwelled areas. The result-

TABLE 29. ESTIMATED FISH PRODUCTION IN THREE OCEAN COMMUNITIES (ADAPTED FROM RYTHER, 1969).

Marine environment	Mean primary productivity (g C/m²/yr)	Trophic levels	Efficiency (%)	Fish production mg C/m²/yr
Oceanic	50	5	10	0·5
Continental shelf	100	3	15	340
Upwelled	300	1·5	20	36 000

ing potential fish production is shown in Table 29. From these results it may be readily concluded that the upwelled areas should produce the largest fisheries and that the continental shelf areas and oceanic are of decreasing importance. This is borne out in part by the fact that the anchovy fishery of Peru is the largest fishery in the world, exceeding the total catch of an intensive fishing nation such as Japan; in contrast very few fisheries are conducted in the open ocean waters except where there is a continental shelf, such as in the northern hemisphere off the Grand Banks. The description of food chains given by Ryther (1969) may be open to certain criticisms as to details of his calculations (e.g. Alverson *et al.* 1970) but the general approach is illustrative of the importance of trophic relationships in assessing the productivity of the ocean.

Riley (1946) first considered quantitative relationships governing productivity at the primary and secondary levels in the marine food chain. Using data from Georges Bank, Riley (1947) considered two approaches; in the first several quantities were measured simultaneously and their interrelationships derived by some statistical method, such as by multiple correlation. In the second approach, a number of simplified assumptions were made, based on experimental data, and these were synthesized into a mathematical description of the events.

As an example of the first approach, Riley (1946) determined multiple regression equations between plant pigments, the depth of the water, temperature, the amount of phosphate, the amount of nitrate, and the abundance of zooplankton. Proportionality 'constants' in the equations varied with season and the effect of changing the mean value for each component by one standard deviation caused different seasonal effects which indicated the temporary importance of each factor. This is illustrated in Table 30 which shows that major changes in the phytoplankton crop were caused by increasing the nitrate concentration during March,

TABLE 30. PERCENTAGE CHANGE IN THE PHYTOPLANKTON CROP PRODUCED BY INCREASING THE VALUE OF EACH ENVIRONMENTAL FACTOR FROM ITS MEAN TO THE LIMIT OF ITS STANDARD DEVIATION (FROM RILEY, 1946).

	Sept.	Jan.	Mar.	Apr.	May	June
Depth	−1	60	−20	11	0	−1
Temperature	−9	−37	−31	−5	10	−26
P	7	74	−23	−24	6	−28
N	7	41	−57	1	−21	1
Zooplankton	−6	5	−1	−9	−31	−10

phosphate concentration during April, and zooplankton during May. Thus it is apparent that no one factor exercises complete control over phytoplankton production and that the relationship between the few major components in Table 30 indicates a complex system in which one factor after another gains momentary dominance. Combining data from different seasons Riley (1946) developed a multiple regression equation in which

$$PP = -153t - 120P - 7 \cdot 3N - 9 \cdot 1Z + 6713 \qquad (89)$$

where PP was the phytoplankton crop in Harvey pigment units (not used today, but proportional to chlorophyll a), t was the temperature, P and N were phosphate and nitrate concentrations, respectively, and Z was the zooplankton standing stock. This equation reproduced the observed data with an average error of about 20% for all seasons.

More recent studies using various types of multiple component analysis have been used to show the principal components governing plankton production in different environments. Williamson (1961) showed that out of eleven factors in a correlation matrix, four accounted for 83% of the variance in plankton abundance in the North Sea and of these, one factor related to the extent of vertical mixing in the water column predominated by accounting for 48% of the variance. Walsh (1971) showed that 66% of phytoplankton variation across the Antarctic Convergence could be predicted in terms of water mass structure, turbulence, light, silicate, and an index of heterotrophic conditions. However, as the latter author has concluded, definite statements about functional relationships between the biological and habitat variables cannot be made because of the inherent non-causal nature of correlation and regression analyses. A similar conclusion lead Riley (1946) to consider a second approach to describing changes in a plankton community. In this approach he assumed that the rate of change in a phytoplankton community could be determined as the difference in reaction rates between processes of accumulation and loss of biomass (or energy) in the plankton population. Riley considered that the most important reactions could be summarized by an equation which can be given in a modified form as

$$\frac{dN}{dt} = N(P_h - R) - G \qquad (90)$$

in which dN/dt was the rate of change in the phytoplankton population (N) having a photosynthetic rate (P_h), a respiration rate (R), and a rate of grazing by zooplankton (G). Riley (1947) considered that each term in the above equation would be subject to environmental influence but that a description of their effect could be given in some mathematical form. Thus the average light intensity ($\bar{I}$) to the depth of the mixed layer, D_m, can be given by the equation,

$$\bar{I} = \frac{I_0}{kD_m}(1 - e^{-kD_m}), \qquad (91)$$

I_0 is the photosynthetic radiation at the surface and k is the extinction coefficient (for the derivation of this equation see Section 3.1.4). The photosynthetic response to this light intensity can then be obtained from a P vs I curve (e.g. Fig. 25) which might be expressed in the form

$$P_h = P_{max}(1 - e^{\alpha(I_c - \bar{I})}) \qquad (92)$$

where P_h is the photosynthetic rate at $\bar{I}$ and P_{max} is the maximum photosynthetic rate for the phytoplankton population, and α is a constant.

Since in eqn. (92) the compensation light intensity (I_c) is determined by the respiration (see Fig. 25), the term R in the eqn. (90) does not have to be treated separately. However, temperature (which was included in Riley's original equation as only affecting respiration) should be included as an effect on the photosynthetic rate. This has been discussed in Section 3.1.3 and using Eppley's empirical relationship between temperature and photosynthetic rate, a temperature adjustment can be included having the form

$$P_h = P_{max}\, e^{(at-b)} \tag{93}$$

where a and b are constants and t is the temperature.

On the basis of more recent findings (see Section 3.1.5) the nutrient limitation on phytoplankton growth can now be expressed in the form of Michaelis–Menten kinetics as

$$P_h = P_{max}\, \frac{[S]}{[S]+K_m} \tag{94}$$

where P_{max} is the maximum rate of photosynthesis at which an increase in the rate limiting nutrient concentration $[S]$ no longer results in an increase in P_h, and K_m is the Michaelis constant for the uptake of the nutrient.

The grazing component, G, can also be expressed more appropriately than in Riley's original equation. Thus if $G = HR'$ where H is the standing stock of zooplankton and R' is the ration per unit time per animal, then from Section 4.2.1

$$R' = R'_{max}(1-e^{-k'N}) \tag{95}$$

where N is the concentration of the phytoplankton population. Thus one attempt to construct a determinate model of phytoplankton productivity on the basis of Riley's classical paper might be written as

$$\frac{dN}{dt} = N\left[P_{max}(1-e^{\alpha(I_c-\bar{I})})\left(\frac{[S]}{[S]+K_m}\right)(e)^{(at-b)}\right] - HR'_{max}(1-e^{-k'N}) \tag{96}$$

where $\bar{I}$ is defined by eqn. (91) and all the variables affect dN/dt independently.

Further embellishments of this type of equation can be made by adding factors, such as might be required to account for the sinking rate of phytoplankton, or by using other expressions for the photosynthetic light response, such as are discussed in Section 3.1.4. In Riley's (1946) original model, which differs from eqn. (96), he was able to reproduce the essential features of the phytoplankton standing stock with approximately the same precision as was obtained by the statistical estimate given in his first approach eqn. (89). Riley (1947) extended his determinate equations to express changes in the herbivorous zooplankton population (dH/dt) as

$$\frac{dH}{dt} = H(gP - R'' - aC - D), \tag{97}$$

where H was the herbivorous zooplankton population, gP was the input from grazing on phytoplankton, R'' was loss from zooplankton respiration, aC was loss by carnivorous predation and D was natural mortality. Many additional descriptive models using both approaches have been developed since Riley's (1946 and 1947) papers. Cassie's (1963) paper serves as a more elaborate example of recent developments in statistical analyses of plankton communities while Patten (1968) has reviewed models which synthesize empirical relationships between the major components governing the production of plankton.

The structure of a plankton community is often not so readily resolved into definite trophic levels as indicated by the use of Riley's models. More often there is a complex food web in which a species of plant or animal may belong to more than one trophic group at any time (e.g. as illustrated in Fig. 48). There are very few quantitative descriptions of such food webs but one has been attempted for the planktonic communities of the Black Sea (Petipa *et al.*, 1970); unfortunately the data presented in this study contain several inconsistencies but the approach serves as an example of a planktonic food web. There are also one or two similar studies in which authors have considered the distribution and flow of energy (or biomass) in other aquatic environments (e.g. the Thames river study by Mann, 1965; the Georgia salt marsh study by Teal, 1962).

In order to understand the energetics of the Black Sea planktonic communities it is necessary first to describe the trophic levels involved, and particular attention is given here to Petipa's description of the epiplanktonic community. The phytoplankton of this community consisted of autotrophic organisms (such as chlorophyll containing diatoms and dino-flagellates) and saprophagus organisms (principally *Noctiluca* which feeds by engulfing particulate food, 70 to 90% of which was reported to be detrital in origin). Thus the authors chose to consider these two groups of organisms as a single trophic level, based on the fact that both groups of organisms were involved in the primary formation of particulate material for filter feeding animals. The distribution of biomass (standing stock) within the phyto-plankton community was over 90% in favour of the saprophagus *Noctiluca*. It appears from the data, however, that this organism had a very high death rate (42% per day) and that this was the chief source of detritus being utilized by filter feeders. In contrast the smaller autotrophic phytoplankton were producing one or two generations per day with a natural mortality of less than 5% per day. Herbivorous organisms formed the second trophic level and these included nauplii of copepods, some copepodite stages, *Oikopleura* and the larvae of some benthic molluscs and polychaetes. Omnivorous organisms, which consumed both plant and animal plankton, formed the third level and consisted mostly of the later stages of the copepods, *Acartia*, *Oithona*, and *Centropages*. One primary carnivore, adult *Oithona*, was recognized as the fourth trophic level, while the fifth and sixth trophic levels were recognized as secondary carnivores (mainly *Sagitta*, which ate both herbivores and primary carnivores) and a tertiary carnivore (*Pleurobrachia*) which fed off all other zooplankton; for most calculations the two latter levels were considered together as one. A diagram of this food web is shown in Fig. 49. While the figure represents fixed relationships described above it was also considered by Petipa *et al.* (1970) that certain organisms transferred from one level to another during their life time so that *Oithona*, for example, transferred to the primary

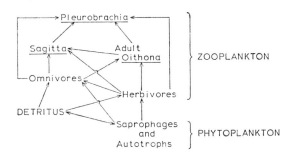

FIG. 49. Food web of the epiplankton community in the Black Sea during early summer (re-drawn from Petipa et al., 1970).

carnivore level on becoming an adult; in addition, a number of early copepodite stages transferred from the herbivore to the omnivore level as they matured.

Petipa *et al.* (1970) considered that the diel rate of production of matter (P') at one trophic level was determined as the number of organisms eaten (G), the number of organisms dying per day (M), the number of organisms transferring to the next level per day (L) and ($B_1 - B_0$) the difference in standing stock at the end of the day. The best units in which to compare the activity of different levels in such a food web are units of energy (e.g. calories/m²/day) although actual measurements are usually made in terms of numbers or biomass of organisms which are assigned caloric equivalents. Experimental data on the six trophic levels described above were collected from an area of high stability in the Black Sea. Food composition and daily rations were obtained from gut contents; reproductive rate of the algae and the respiration of animals were obtained from samples incubated *in situ* and the total weight increment of animals was determined from the duration of stages at different temperatures and the known weight of each stage. The results of these measurements are shown in Table 31. From these data it is apparent that the phytoplankton community is predominated

TABLE 31. PRODUCTION OF THE TROPHIC LEVELS OF BLACK SEA EPIPLANKTON COMMUNITY (mg/m²/DAY) FROM PETIPA *et al.* (1970).

Trophic Level	Amount of food eaten from a given level (G)	Death rate (M)	Difference between final and initial standing stock ($B_1 - B_0$)	Transport to the following level (L)	Production (P)	Standing stock of living organisms (B)	P'/B (%)
Primary producers and saprophages	478·5	12 593·8	5397·2	—	18 469·5	29 249·6	63
Herbivores	361·2	4·0	−6·3	156·0	515·0	792·0	65
Omnivores	67·0	1·2	20·0	82·0	170·4	193·5	88
Primary carnivores	53·5	1·3	27·4	—	82·2	172·2	47
Secondary-tertiary Carnivores	33·0	2·8	−1·1	—	34·7	560·5	6·2

by the rather unusual death rate which has been attributed to the saprophage, *Noctiluca*. Further it is apparent that both biomass and production show a general decrease going from the lowest trophic level towards the highest. Production per unit biomass (P'/B) would have been highest at the primary level if data on *Noctiluca* had been excluded. There is some evidence in the data that P'/B ratios decrease with increasing trophic levels, as might be expected for slower growing organisms which have to spend an increasing amount of energy on capturing their prey.

Dickie and Mann (1972) have attempted to summarize P'/B ratios for different trophic levels and their results are reproduced in Fig. 50. P'/B ratios in Table 31 are not directly comparable to data in Fig. 50 since the former are based on daily estimates and do not allow for an organism's life cycle which may impose a natural restriction on its growth for part of a year. Thus phytoplankton may continue to multiply throughout the year while more complex organisms have to grow to maturity, mate and reproduce; these processes

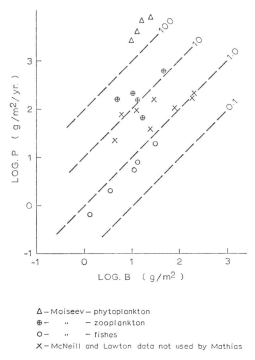

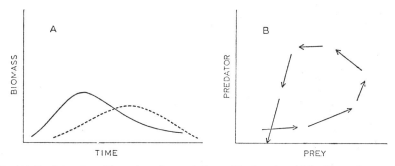

Δ – Moiseev – phytoplankton
⊕ –   "   – zooplankton
○ –   "   – fishes
✕ – McNeill and Lawton data not used by Mathias

FIG. 50. Production (*P*) and biomass (*B*) relationships (data accumulated by Dickie and Mann, 1972 from various sources. Dashed lines indicate log. intervals of constant *P/B*).

may require several years in higher organisms, including many fish. P'/B ratios on an annual basis indicate that at the three principal trophic levels, the respective ratios are *ca* 300 for phytoplankton, 10 to 40 for most zooplankton and 1 or less for most fishes depending on how long they take to complete their life cycle. The data also indicate that the annual average standing stock at different trophic levels may be very similar, but that low ecological efficiencies reduce the transfer of high productivity at low trophic levels, as material is moved up the food chain.

An important property of a food web is its stability. The simplest case is to consider a single predator and a single prey over the course of time (Fig. 51). If there are not other restrictions on this system the predator will increase to a point where it runs out of food and

FIG. 51. (A) Predator (– – –)/prey (———) association with time in which (B) predator exhausts its food supply and the population is extinguished (redrawn from Maly, 1969).

then declines and become extinguished. Such a system lacks stability. Under natural conditions, however, various restrictions are placed on predator/prey associations which tend
to dampen out violent oscillations. Dunbar (1960) suggested that in general oscillations are
bad for any system and that violent oscillations may be lethal. This has lead to a definition
(Hurd *et al.* 1971) in which it can be stated that stability is the ability of a system to maintain
itself after a small external perturbation.

Factors which tend to stabilize natural systems have been discussed by several authors
(e.g. Rosenzweig and MacArthur, 1963; McAllister *et al.*, 1972). MacArthur (1955) and
others have postulated that development of many species (i.e. a high diversity) is the principal component in establishing community stability. Since more energy may be required in
establishing a more complex food web (e.g. see Ryther's three trophic relationships) it is
also apparent that where stability has been acquired through diversity, productivity per unit
biomass will be low. Thus in general, tropical plankton communities and food webs tend to
have a high diversity, low productivity, and high stability. In contrast a temperate plankton
community has a high productivity caused by one or two species of zooplankton (e.g.
Calanus plumchrus and *cristatus* in the subarctic Pacific) and a low stability in which there
are large seasonal fluctuations in production. Some other important factors leading to the
stabilization of a system are the imposition of a limitation on the predator, other than its
supply of prey; a hiding place for the prey; periodic migration of the predator away from
its food source; a threshold concentration below which the prey is not consumed by the
predator; and, the patchiness of prey distributions. Examples of these stabilizing influences
can be found in the aquatic environment. Thus the diel migration of zooplankton into the
euphotic zone may be interpreted as adding to community stability (as well as being of some
physiological advantage to the zooplankton population, e.g. see McLaren, 1963). The
possible occurrence of a threshold prey concentration in phytoplankton/zooplankton
relationships (e.g. Parsons *et al.* 1967) and the ability of zooplankton, such as euphausiids,
to change from being carnivores to herbivores (Parsons and LeBrasseur, 1970) are further
examples of ways in which stability of a community may be maintained. On the other hand,
instability may be imparted on a plankton community by time delays (Dickie and Mann,
1972); one example of this is the maintenance of the barnacle community in the Firth of
Clyde which is highly dependent on the timing of the *Skeletonema* bloom in association
with the release of barnacle larvae (Barnes, 1956).

An important observation regarding the stability of a system is based on the natural
tendency for components of a system to be more variable than the system itself. Thus it is
apparent from nearly all ecological oceanographic data (e.g. Riley's 1947 data on George's
Bank) that individual components of a food web leading to production at a particular trophic
level are more variable *in toto* than the variability in the production level itself. This has
been expressed by Weiss (1969) and given by Dickie and Mann (1972) as a definition: viz. a
system exhibits stability when the variance of the whole is less than the variance of the
parts. This definition has some merit in that it can be expressed mathematically as

$$S = \frac{N_i V}{\Sigma(v_1 + v_2 + v_3 \dots v_i)} \tag{98}$$

where S in an *index* of stability of a system having a variance V and a number of components
(N_i) making up the system with variances of v_1, v_2, $v_3 \dots v_i$. An advantage to the use of
the above ratio as an index of stability is that it removes the problem of deciding on the
relative size of perturbations as destabilizing effects on a system. Thus one system which

shows regular oscillations over time of several hundred percent may be as stable as a system which shows practically no variation with time.

Patten (1961 and 1962b) considered the stability of a plankton community and its environmental parameters. The stability index evolved by Patten (1962b) was expressed as

$$S' = \frac{\sum\limits_{j=1}^{m} \det P_j}{\sum\limits_{j=1}^{m} (s/\bar{x})_j} \tag{99}$$

where

$$P_j = \begin{bmatrix} P_{id} & P_{ii} \\ P_{dd} & P_{di} \end{bmatrix} \tag{100}$$

which is the matrix of transition probabilities for the jth of m variables, P_{id} being the probability for a decrease in value of the variable following an increase, P_{ii} that of an increase following an increase and so on. det P_j is then determined as

$$\det P_j = P_{id}P_{di} - P_{ii}P_{dd} \tag{101}$$

and $(s/\bar{x})_j$ is the standard deviation divided by the mean of the jth variable. Equation (99) takes into account both the direction of changes and their amplitude. Thus if a parameter tends to show more of an increase than a decrease (or vice versa) the stability index will be negative; a large $s/\bar{x}$ ratio will further increase the value S' indicating a lack of stability. In comparing S' for components of the physical environment and plankton community, Patten found that the plankton (S'_p) was 6·7 times more stable than the environment (S'_e); this is not unreasonable since the plankton must to some extent absorb the 'shock' of environmental change and if $S'_p < S'_e$ the plankton community would collapse.

From these observations it is apparent that variability in the components of a system contribute to the systems stability which is the opposite process to that contributing to the instability of a machine (Weiss, 1969), or a highly controlled ecosystem, such as is attempted in aquacultural projects. In controlled operations, such as machines, the variance of each of the components add up to describe the variance in the performance of the total unit. Only by imposing maximum control over the components is it possible to predict the operation of the machine. In contrast, in systems the components vary widely but the system itself (if stable) may show only small (or if large, regular) oscillations. This contrast is brought out when attempts are made to simulate a system using a computer model which inadequately describes the buffering action of the natural environment. Thus small changes in a component of such models can be shown to produce practically any change in a population, however unrealistic such changes may be in a real ecosystem (e.g. McAllister, 1970). In contrast it appears that a natural aquatic ecosystem is made up of highly variable components which at some point in their oscillating states, interlock to allow for a flow of energy up the food chain; the exact position of the interlocking point in any process being maintained by fluctuations about a mean, rather than by rigid control of an absolute value. One corollary of these observations is that a perturbation applied to the top of a food chain should have more effect on the nature of the food chain than if a perturbation of similar magnitude is applied to the bottom of the food chain. This may be seen in part by two large scale perturbations applied to sockeye producing lakes in British Columbia. In the first experiments (Foerster and Ricker, 1941; Foerster, 1968) 90% of the sockeye predators were removed from a lake with a resulting *ca* 300% increase in the biomass of sockeye produced in the lake. In a second experiment (Parsons *et al.*, 1972; Barraclough and Robinson, 1972) nutrients added to a large sockeye-producing lake were increased by 100% which resulted in only about a 30% increase in the biomass of sockeye, due to depensatory components (particularly a temperature effect). While these two experiments are only very crudely comparable they serve to illustrate that the proportional increase in fish production was at least twice as effective when a perturbation was applied to the top, in contrast to the bottom, of an aquatic food chain. However, while production is the only consideration in this section, the removal of a predator as a means of raising production may not be economically sound if the same predator is the focal point of another interest, such as the sport fisherman.

Other work on lake communities has been helpful in obtaining an understanding of marine pelagic food chains. Since lakes are relatively discrete bodies of water, it is often easier to establish cause and effect relationships in the aquatic food web than in the sea where the exchange of water may remove organisms from the study area. Density dependent relationships between the growth of planktivorous fish, and phytoplankton and zooplankton growth rates and abundance are particularly difficult to establish in the marine environment. In lakes, however, it can be demonstrated (e.g. Brocksen *et al.* 1970) that the abundance of fish is generally related to the abundance of plankton. However, depensatory mechanisms are often apparent; thus in the case of sockeye salmon producing lakes, there is a general relationship between zooplankton production and sockeye salmon production *between different* lakes. However, *within* a single lake, a large number of young fish may compete so heavily for a limited amount of food that the entire population is weakened and the number of fish surviving to become adults may be very much smaller than if fewer young fish were present initially. Another depensatory mechanism has been discussed by Kerr and Martin (1970). In a study of trout populations in Ontario lakes, these authors found that while there was a general relationship between phytoplankton productivity and trout production there was no relationship between trout production and the length of the food chain in different lakes, as implied by the general relationship in eqn. (88). Thus the biomass of large trout feeding as piscivores was similar to the biomass of small trout feeding as planktivores (other factors being equal, between lakes). This was explained by the fact that the large piscivorous trout expended less energy on feeding and were more efficient feeders than the smaller, planktivorous trout. While this depensatory mechanism may be valid in a relatively confined environment, it is probably less valid in marine ecosystems where herbivorous zooplankton are much larger and often more densely aggregated than in lakes.

Predation by planktivorous fish on the plankton community may also affect the structure of an aquatic food web. This has been illustrated by Brooks and Dodson (1965) in the case of a number of lake communities in the eastern United States. Under conditions in which there were few planktivorous fish, it was found that large zooplankton generally predominated. It was assumed that this was because large herbivorous zooplankton are generally more efficient phytoplankton feeders since they filter both large and small phytoplankton. However, in lakes where planktivorous fish were predominant, large zooplankton were selectively grazed by the fish; this markedly decreased the dominance of large zooplankton, allowing smaller zooplankton to flourish. Further, since the smaller zooplankton could not feed off large phytoplankton, the latter also tended to flourish. Thus the whole size structure of planktonic organisms in similar lakes was largely determined by the abundance of planktivorous fish.

CHAPTER 5

BIOLOGICAL CYCLES

IN THE preceding sections it is apparent that organic and inorganic materials are distributed throughout the food chain and that the mineralization of organic matter back to carbon dioxide and inorganic radicals serves, in itself, to assure a 'food' supply for autotrophic and chemosynthetic organisms. Thus the distribution of carbon and other elements tends to follow a cyclic pattern in which a single element may be present in several different phases (e.g. as a solid, gas, or in solution). In contrast to the recycling of materials, energy can only be used once. Mann (1969) has commented that while materials circulate through the biosphere, energy flows through the system in one direction, entering as light during photosynthesis and being lost as heat during respiration. In one sense this is true but the passage of energy through organic compounds into inorganic compounds (e.g. H_2S, NH_3^+, etc.) and back to organic compounds and the food chain, represents a redistribution of energy which is cyclic.

Many attempts have been made to depict organic and inorganic cycles in aquatic systems (e.g. Strickland, 1965; Kuznetsov, 1968; Nakajima and Nishizawa, 1968; Riley and Chester, 1971). The closer these come to representing all the intricate possibilities involved in the flow and feedback of materials, the more complex they become. Unfortunately, this leads to difficulties in understanding their significance. In an attempt to clarify this situation, we have represented degrees of cyclic complexity by separating overall processes from some of the more detailed interdependent reactions.

5.1 ORGANIC CARBON AND ENERGY CYCLES

In a preceding section it has been noted (Section 2.4) that the largest fraction of organic carbon in the oceans exists as debris, either as organic compounds dissolved in sea water or as particulate organic detritus. It is also apparent from the utilization of organic substrates by heterotrophic organisms, as well as from the utilization of particulate detritus by zooplankton (e.g. Chindonova, 1959), that the reservoir of organic debris in the sea feeds back into the food chain. From these observations Riley (1963) has concluded that there is a flexible system of reversible reactions allowing organisms to draw upon the reservoir of organic carbon and replenish it in a variety of ways. This in turn has tended to stabilize the aquatic environment by providing a food source for living organisms over a longer period of time than that in which a single phytoplankton bloom can be sustained by the environment.

This system can be illustrated by Fig. 52, which has been numbered to show (1) the input

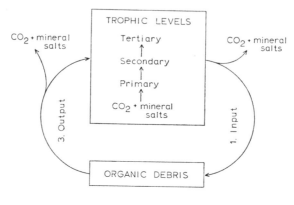

FIG. 52. A generalized organic carbon cycle showing the exchange of carbon between the food chain and the organic debris of the sea.

into the system from the food chain, (2) the reservoir of organic debris which is assumed to have reached a steady state over a long period of time, and (3) the output from the reservoir which includes both the mineralization of organic matter to carbon dioxide and inorganic radicals, as well as the utilization of organic matter by the food chain. Olson (1963) has discussed the kinetics governing the input of organic material (L) to a standing stock of organic carbon (X). In the simplest case of a continual input and with a constant decomposition rate, $k(\text{time}^{-1})$, changes in X with time can be given by the equation

$$X = \frac{L}{k}(1 - e^{-kt}).\tag{102}$$

In a situation where sufficient time has been allowed for the system to come to equilibrium (t large), a steady state value of X will be reached such that

$$X_{\text{ss}} = \frac{L}{k}.\tag{103}$$

Minderman (1968) has criticized the use of this approach in terrestrial environments on the grounds that the initial decay of fresh organic material is much faster than can be accounted for by use of the kinetics given above (see also Section 2.4). In aquatic environments, however, decomposition starts as the debris descends towards its final place of accumulation; in shallow seas this will generally be the sediments while in the deep oceans the largest accumulation of organic material is in the deep water column. Thus the initial rapid loss of organic material will probably have already occurred when the input term (L) is assessed. Skopintsev (1966) using the above kinetics [eqn. (102)] estimated that the one or two parts per million of soluble organic carbon in deep ocean waters had accumulated over a period of several thousand years; a value which is similar to the residence time of 3400 years found by radiocarbon dating (Williams *et al.*, 1969). Using the same approach, the turnover time of organic carbon in a near shore sediment was found to be about 30 years (Seki *et al.*, 1968).

The need to consider biological systems in terms of energy units instead of units of weight, has already been expressed at several points in this text. In practice, however, the expression of results in terms of weight per unit area is more easily conceived and compared with everyday problems of agriculture, fisheries, and nutrition. However, the absolute need to use energy units when discussing biological cycles emerges whenever a consideration is given to autotrophic and chemosynthetic processes in an environment. This is illustrated in Fig. 53, which is modified from Sorokin (1969).

The left hand side of the figure is divided into aerobic and anaerobic environments; since oxygen disappears in anaerobic environments, the best division of the two environments

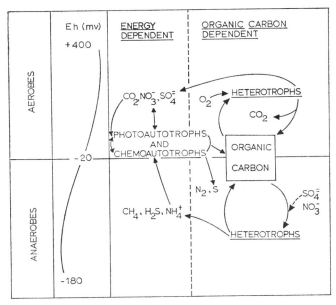

FIG. 53. Organic carbon and energy dependent cycles in marine aerobic and anaerobic environments (modified from Sorokin, 1969).

is in terms of the 'oxidizing potential', or E_h, measured in millivolts. Generally most of the environments discussed in this text have been aerobic environments in which the oxygen concentration was adequate to supply all trophic levels. In the Black Sea, and in some fjords and lakes, however, anaerobic environments exist below the surface and it is at the interface of these two zones that a number of biological reactions may occur. The areas mentioned above are not large compared with the hydrosphere but it should be considered that certain near-shore sediments, which are not mixed by wave action, may contain very similar aerobic/anaerobic zones. The important aspect of an anaerobic zone wherever it occurs is that it represents a storehouse of chemical energy which may have some ability to feed back into the food chain; certainly this has already been demonstrated in meromictic lakes and the Black Sea (Sorokin, 1969). In these environments different groups of bacteria exist which can decompose organic material by using sulphate and nitrate as a source of oxygen, and in the process form reduced substances, such as CH_4, H_2S, and NH_4^+. The latter compounds can be utilized by other bacteria, some of which are strictly chemoautotrophs since they can use CO_2 as a source of carbon, and inorganic compounds as a source of energy. The extent to which CO_2 is taken up independently of photosynthesis has been studied by a number of authors (e.g. Romanenko, 1964a, b; Sorokin, 1969) and a discussion of these processes is given in Section 3.1.6.

The top of Fig. 53 is divided into energy-dependent reactions and organic carbon-dependent reactions. In moving from either photoautotrophic or chemoautotrophic reactions, the cycle can either continue to be described in energy units, or quantities of organic carbon can be substituted (the latter are generally easier to measure in small amounts). Thus a photosynthetic organism receives its energy to grow from light, but once new organic material is formed, it can be expressed as so many calories of organic matter or in terms of dry weight of organic carbon, wet weight and so on. All other steps in the food chain, represented by the "organic carbon" box in Fig. 53, may also be expressed in terms of biomass. However, due to the very different inorganic chemical composition of marine organisms (including the water and ash content of plankton, e.g. Table 12), it is often advisable to retain units which express either organic carbon or a growth-limiting element, such as nitrogen, and to avoid units such as total wet or dry weight. Some authors recommend the exclusive use of energy units but it is apparent from Table 11 that the primary producers may at times be rich in energy compounds (i.e. carbohydrates and lipids) and low in nitrogen compounds (i.e. proteins) and vice versa depending on the availability of inorganic nitrogen. Thus two different metabolic cycles or food webs may exist in nitrogen limited (e.g. tropical) and nitrogen available (e.g. temperate) environments; the former being characterized by excessive amounts of energy rich compounds and the latter being an energy starved but nitrogen rich system.

Figure 37, in Section 3.2.1, is another representation of a carbon cycle; in this figure emphasis is given to the loss of organic material through various processes in the marine food chain. A large component which could be added to this cycle is the loss of materials to the benthic community by sinking. In certain areas, such as the Grand Banks, an important pathway for organic material is through sedimentation of phytodetritus, benthic filter feeders and from thence to demersal fish which inhabit such shallow areas in great abundance.

5.2 INORGANIC CYCLES

Many of the earliest studies on the inorganic micronutrients of the sea were concerned with the study of the phosphorus cycle (e.g. Harvey, 1957, and references cited therein). The occurrence of phosphorus in three different forms (i.e. particulate phosphorus, soluble organic phosphorus, and inorganic phosphate) over a period of a year in a coastal environment is shown in Fig. 20, Section 2.1. From this figure it is apparent that the biological phosphorus cycle of the sea involves the uptake of inorganic phosphate by phytoplankton during the summer in temperate latitudes. Phosphorus is then redistributed as particulate and soluble organic phosphorus, the latter resulting from the breakdown of cellular material, as well as from the release of organic phosphorus from living plants and animals. In tropical and subtropical latitudes seasonal cycles are far less well defined and the distribution of phosphorus may be assumed to be in a state of flux in which phosphorus from decomposition processes is utilized as it becomes available.

Pomeroy *et al.* (1963) and Johannes (1965) showed that zooplankton, and particularly marine Protozoa, could excrete phosphorus as organic compounds and as inorganic phosphate, in daily amounts in excess of their body phosphorus content. The excretion of phosphorus by zooplankton was shown by Martin (1968) to be inversely proportional to food abundance; the explanation for this was that phosphorus was used for storage products or egg production when food was abundant, but excreted due to a relative increase in body metabolism when food was scarce. A distribution of dietary phosphorus for *Calanus* during active feeding (April) has been given by Butler *et al.* (1970) as 17·2% retained for growth, 23·0% excreted as fecal pellets and 59·8% excreted as soluble phosphorus. Thus the grazing activity of *Calanus* effectively returns more than 80% of the phytoplankton phosphorus to the environment. This process of phosphorus remineralization has been observed in nature. For example Cushing (1964) showed that during 10 weeks of the spring phytoplankton/zooplankton bloom in the North Sea, inorganic phosphate did not decrease below 0·6 μg at/l. However, Antia *et al.* (1963) observed a decrease in phosphate to 0·1 μg at/l after two weeks of phytoplankton growth in the absence of zooplankton.

Both the excretion or organic phosphorus by phytoplankton and the uptake of phosphorus from organic phosphorus compounds for phytoplankton growth have been demonstrated by a number of authors (e.g. Kuenzler, 1965 and 1970). Studies on the remineralization of phosphorus from decaying phytoplankton in the absence of light (Antia *et al.*, 1963) showed that half the organic phosphorus could be released as reactive inorganic phosphate in a period of two weeks. Phosphorus lost to sediments through sinking as particulate phosphorus may be partly converted into insoluble mineral phosphates or recycled by the benthic animals and microflora.

A general summary of the major pathways in the phosphorus cycle of the sea is shown in Fig. 54. From this cycle and the observations given above it is apparent that the various

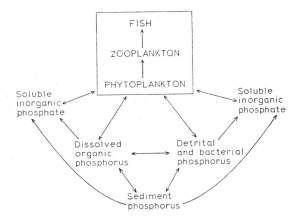

FIG. 54. Major pathways in the phosphorus cycle of the sea.

orms of phosphorus are often readily exchangeable as metabolites. The rate of turnover, fespecially in the presence of zooplankton, indicates that phosphorus is generally available in the marine environment although the absolute concentration of phosphate may sometimes be sufficiently low as to determine the actual growth rate of the phytoplankton species.

Phosphorus is readily hydrolysed from organic compounds, either by hydrolysis at the alkaline pH of sea water or by phosphatases, which are hydrolytic enzymes present in many bacteria and on the surface of some phytoplankton, particularly those from environments low in inorganic phosphate. In contrast, organic nitrogen is cycled with much greater difficulty since its fixation, reduction, and oxidation back to nitrate requires the exchange of energy. Thus in the same experiment in which half the phytoplankton phosphorus was mineralized in 14 days, no mineralization of organic nitrogen was observed in a period of up to 75 days (Antia *et al.*, 1963). However, as in the case of phosphorus, zooplankton feeding greatly enhances the recycling of nitrogen, either as ammonia (e.g. Harris, 1959; Corner and Newell, 1967) or through the release of soluble organic nitrogen compounds (Johannes and Webb, 1965; Webb and Johannes, 1967 and 1969). A dietary nitrogen budget for actively feeding *Calanus* has been given by Butler *et al.*, (1970) as 26·8% retained for growth, 37·5% excreted as fecal pellets, and 35·7 excreted as soluble nitrogen compounds.

In temperate waters the uptake of nitrate nitrogen follows a seasonal cycle similar to that for phosphate with the exception that the supply of nitrate in the surface layers often becomes completely exhausted during the summer months, while phosphate may continue to be present in low concentrations. In tropical and subtropical waters nitrogen availability is often regarded as the rate-limiting nutrient throughout the year. Dugdale and Goering (1967) have shown that ammonia is taken up by phytoplankton more rapidly than nitrate; this was particularly noticeable in phytoplankton from subtropical compared with temperate regions. Urea may also serve as a nitrogen source for some phytoplankton and significant concentrations (*ca* to 5 μg at/l) of this compound have been found in coastal and oceanic waters (Newell, 1967; McCarthy, 1970; Remsen, 1971). In tropical environments the fixation of molecular nitrogen by certain bacteria and more especially by blue–green algae, may be very important. Goering *et al.*, (1966) measured the uptake of molecular nitrogen in waters containing *Trichodesmium* sp. and found a maximum rate of 0·32 μg N/l/hr. The loss of nitrogen through denitrification of nitrate appears to be an anaerobic reaction. However in some aerobic environments, denitrification may occur in the presence of detrital particles; this is probably due to the formation of anaerobic microzones of bacterial activity within the particles (Jannasch, 1960).

The conversion of ammonia to nitrite and nitrate may proceed in any environment by a series of reversible reactions. These are largely carried out by bacteria but some phytoplankton may produce extracellular nitrite during the course of nitrate uptake. A general schematic summary of nitrogen conversions in the sea has been given by Vaccaro and Ryther (1959) and a modified version of their diagram is presented in Fig. 55. In this diagram the food chain, as the primary consumer of inorganic nitrogen, has been represented at the top, and the bacteria, as the primary agents in recycling nitrogen are represented at the bottom. However, as in the case of phosphorus, nitrogen can be effectively recycled between

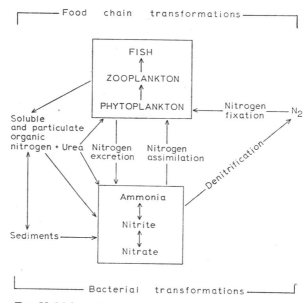

FIG. 55. Major pathways in the nitrogen cycle of the sea.

phytoplankton and zooplankton under conditions of a plankton bloom which favours both zooplankton grazing and the growth of phytoplankton.

Ammonia, particulate organic nitrogen, and soluble organic nitrogen from the food chain may be present in different proportions in different environments; the bacterial oxidation products of these reduced forms of nitrogen may also occur in different proportions, especially in depth profiles. In particular, the occurrence of a nitrite maximum at the bottom of the euphotic zone is often a well-defined phenomenon in thermally stratified surface waters. Nitrite can be produced either through the oxidation of ammonia (a dark reaction) or by reduction of nitrate (stimulated by light). It is possible that the ammonia pathway is carried out largely by bacteria while the nitrate pathway is mediated through phytoplankton. The presence of the nitrite maximum has recently been reviewed by Wada and Hattori (1971), who considered that it was due to nitrite production from ammonia and nitrate at a rate in excess of biological and physical processes which remove nitrite. The latter would include the biological uptake of nitrite precursors (i.e. nitrate or ammonia) in the euphotic zone above the nitrite maximum, and decreased nitrite concentration due to advective processes, particularly below the thermocline.

The cycling of other micronutrients in sea water is generally less well defined than in the case of nitrogen and phosphorus. Silicon, which is required in diatom metabolism for the formation of cell walls, follows a seasonal cycle of abundance in temperature waters. The lack of soluble silicate may to some extent determine species succession from a diatom to a flagellate community. Many biologically essential inorganic materials in sea water are present in very large amounts so that any cyclic process involved in their distribution is not important to the rate control of metabolic processes in the food chain. These nutrients include sodium, magnesium, potassium, calcium, sulphate, water, and carbon dioxide. In some cases elements may become limiting to phytoplankton growth due to their availability and not to their absolute concentration. Thus iron, particularly in sub-tropical and tropical oceanic waters, may become growth limiting in the absence of organic chelators, which makes it available for uptake by plants.

5.3 TRANSFER OF ORGANIC COMPOUNDS WITHIN THE FOOD CHAIN

Certain organic compounds are transferred within the food chain of the sea; this may also involve a cyclic process in which the compounds are recycled between different trophic levels but at present this has not been a subject of investigation. In the following section some examples are given of compounds which are transferred intact between different trophic levels.

The importance of preformed compounds in the nutrition of lower organisms has been well recognized in the vitamin requirements of certain algae. In a review of this subject Provasoli (1958) showed that thiamine, biotin and vitamin B_{12} are often required by marine algae. Since these compounds are also required in a preformed state by higher organisms it is reasonable to consider their transfer up the food chain as a necessary part of food chain stability. The exact state of the preformed compound has been investigated in the case of vitamin B_{12} by Droop et al. (1959), who showed that various analogues and precursors of vitamin B_{12} could also serve to meet the requirement for this vitamin in some algae.

The transfer of an antibacterial substance in the food chain of penguins has been investigated by Sieburth (1961). The compound which was found to have strong antibacterial properties was acrylic acid; this is present in relatively large amounts in a common antarctic phytoplankton, *Phaeocystis* (a genus which is also found in the northern hemisphere). In the antarctic this organism forms substantial blooms which are consumed by *Euphausia superba* or krill; the euphausiids in turn are one of the principal food organisms of penguins. Acrylic acid is transferred to the penguin where it causes bacteriological sterility in the anterior segments of the gastrointestinal tract.

The transfer of fatty acids up the food chain from phytoplankton to fish has been indicated by Williams (1965) and studied experimentally by Kayama (1964) and Lee *et al.* (1971). Kayama (1964) fed a diatom, *Chaetoceros* sp., to brine shrimp which were then fed to guppies. Fatty acid composition of the lipids from the three trophic levels were then examined and the results for C_{12} to C_{22} fatty acids are shown in Table 32. These results

TABLE 32. FATTY ACID COMPOSITION OF *Chaetoceros*, *Artemia*, AND GUPPY OILS (WEIGHT PERCENT OF TOTAL ESTER). [FROM KAYAMA (1964)].

Fatty acid	*Chaetoceros*	*Artemia*	Guppy	
			17±1 °C	24±1.5 °C
Shorter chain	trace	0.4	trace	
12 : 0	0.4	trace	0.2	trace
13 : 0	0.7	trace	trace	trace
14 : 0	13.0	4.8	1.5	0.9
15 : 0	1.8	1.5	trace	0.2
14 : 2	0.6	trace	0.6	0.5
16 : 0	18.1	11.6	22.9	36.0
16 : 1	47.9	44.9	15.9	8.9
16 : 2	2.7	trace	0.2	0.2
16 : 3?	4.0	1.7		0.6
16 : 4?	trace			0.5
18 : 0	0.5	1.9	8.2	9.8
18 : 1	8.7	18.4	18.3	15.0
18 : 2	1.7	0.7	trace	trace
18 : 3	trace	0.5	1.4	0.8
20 : 1		0.9		
18 : 4 & 20 : 2		0.8	0.3	trace
20 : 3			0.2	trace
20 : 4		trace	2.0	2.0
20 : 5		12.0	4.8	4.6
22 : 4			1.3	1.0
22 : 5			6.1	7.3
22 : 6			16.5	11.5

show a transfer of 14:0, 16:0, 16:1, and 18:1 fatty acids, but an apparent synthesis of polyunsaturated fatty acids by *Artemia* (20:5) and guppies (22:5 and 22:6). However, from Table 21 it is also apparent that C_{20} and C_{22} polyunsaturated fatty acids occur among some phytoplankton, in which case they are probably transferred intact up the food chain (e.g. Lee *et al.*, 1971). Hinchcliffe and Riley (1972) fed different phytoplankton diets to brine shrimp (*Artemia*) and found that the levels of saturated acids were comparatively constant regardless of the food organism used. They showed also that polyunsaturated fatty acids occurred in their different phytoplankton diets and that they were transferred to the brine shrimp. However some fatty acids occurred in higher proportions in the brine shrimp than in the algal food (e.g. oleic acid, as shown also by Kayama, 1964). Ackman *et al.*, (1970) studied the fatty acids and lipids in north Atlantic euphausiids and showed differences between samples which could be explained in terms of diet. As a general conclusion, therefore, it appears that under natural conditions, where animals feed on a broad range of food, there will be qualitative similarities but some quantitative differences between the fatty acid composition of aquatic animals and their diet.

From studies on hydrocarbons in algae it appears that an important biological substance found in many algae is a 21 : 6 hydrocarbon, all-*cis*-3, 6, 9, 12, 15, 18-heneicosahexaene or "HEH". This substance is present to the extent of a few thousandths of a percent of the cell weight of algae and is also found in small quantities in zooplankton and higher organisms including oysters, herring and the basking shark, among animals investigated (Blumer *et al.*, 1970). It has been suggested (Youngblood *et al.*, 1971) that this compound is involved in some way in a biochemical function of the reproductive cycle in both marine plants and animals. The evidence for this is at present circumstantial but it appears that polyunsaturated hydrocarbons are highest in algae during exponential growth, they are concentrated in the reproductive structure of at least one benthic algae, and in the specific case of HEH there is some correlation between the predominance of HEH to other hydrocarbons in algal foods and the relative proportions of male animals hatched from *Calanus helgolandicus* when fed different algae. The general chemistry of hydrocarbons in the marine food chain also appears interesting from the fact that different species of algae appear to have quite specific hydrocarbon ratios and that these differ from those synthesized by copepods and present in mineral oils. Zooplankton contain complex mixtures of C_{19} and C_{20} isoprenoid alkanes and alkenes while mineral oils contain many isomers including large amounts of cyclic compounds, but no olefins. Since hydrocarbons are relatively indestructible in the food chain of the sea, it is apparent that transfer routes within the food web as well as the source of oil pollutants might be discovered through hydrocarbon chemistry (Blumer *et al.*, 1971).

CHAPTER 6

SOME PRACTICAL PROBLEMS IN
BIOLOGICAL OCEANOGRAPHY

BEFORE the turn of the last century the earliest publications in biological oceanography were primarily concerned with the taxonomy and distribution of marine planktonic organisms. With the introduction of extensive nutrient analyses in the 1930's, biological oceanography entered a more quantitative era and some efforts were soon made to use biological oceanographic information for the solution of practical problems. Early attempts to relate nutrient distributions to fisheries were not successful but the exercise served to demonstrate where more knowledge was required. Riley's (1946) work on modelling the plankton community lead the way to a much greater understanding of integrated processes. The impediment to further development of models then lay in the inability to collect sufficient data over an area and time scale applicable to the models. This problem is being solved with the use of automated equipment such as autoanalyzers coupled to shipboard computers (e.g. Walsh and Dugdale, 1971). Throughout this brief history of development, however, it has been continually necessary to summarize relationships between plankton and the environment in a variety of empirical expressions, many of which have been discussed in earlier sections of this text. These expressions, (and future modifications) are the building blocks on which practical solutions to biological oceanographic problems must be attempted. While it is recognized that more knowledge would be helpful, solutions to many of today's problems are required today, and where answers are required they should be given to the best of the state of present day knowledge.

In the following section some attempt has been made to identify a number of problems for which solutions may be attempted from a knowledge of biological oceanographic processes. The actual application of biological oceanographic information to the solution of a problem will depend heavily on the local environment and for this reason the following subjects are only considered by example, or from a point of view of general relationships.

6.1 EXAMPLE PROBLEMS IN POLLUTION AND
WATER MASS IDENTIFICATION

The survival of terrestrial bacteria in the marine environment has received considerable aentitton because of man's tendency to regard the sea as a septic tank. Thus one of the commonest forms of coastal pollution is the contamination of beaches and nearshore fisheries with pathogenic organisms from human excrement. Jones (1971) has estimated that 4.4×10^7 lbs of faeces are released per day in the United States and that an appreciable proportion of this is disposed of in the nearshore environment. While the distribution of this material on entering the sea is a problem in physical oceanography, the survival of pathogenic bacteria in sea water is a biological problem. The subject has recently been reviewed by Mitchell (1968) and Jones (1971). Of the many bacteria entering the sea, the presence of *Escherichia coli* is usually taken as an index of sewage pollution. The organism is not itself a pathogen but its presence in large numbers may indicate the presence of pathogenic bacteria and viruses. The actual 'acceptable' level of coliform contamination may vary depending on national standards and use of surrounding beach areas; generally a level of less than 100 coliforms/100 ml in 90% of the samples averaged over a 30-day period is considered an acceptable limit while a level of 1000 coliforms/100 ml would be considered contaminated.

Several mechanisms are considered to cause the destruction of terrestrial bacteria in the sea. The production of antibacterial substances by marine plankton (e.g. Duff *et al.*, 1966; Burkholder *et al.*, 1960) has been shown to include substances active against *Staphylococcus* although only about 40% of the algal species tested by Burkholder *et al.*, (1960) displayed activity against this pathogen. A specific antibacterial activity of the phytoplankter, *Phaeocystis*, was identified by Sieburth (1960) as acrylic acid. In addition to the antibacterial activity of marine phytoplankton it has also been suggested that sunlight, temperature, and specific bacteriophages may cause mortality of terrestrial bacteria in seawater. However, most of these mechanisms might be considered to be equally effective against the indigenous species of marine bacteria and not specific agents in causing the high mortality of terrestrial species. A mechanism which is more specifically related to a difference in properties between freshwater and seawater is the salt content of the latter; salt (NaCl) in itself, however, is not considered to be the lethal agent since several authors have shown that some terrestrial bacteria can tolerate (but not necessarily grow) in NaCl concentrations higher than are found in the sea (Korinek, 1927; Burke and Baird, 1931). On the other hand Jones (1963) found that *E. coli* was killed by natural seawater; a process which could be prevented by the addition of a chelating agent. Jones (1963, 1964 and 1967) concluded that it was the heavy metal content of seawater together with the low concentration of natural organic chelators which was lethal to *E. coli*, and other terrestrial bacteria. Under these circumstances it is apparent that nearshore environments which contain a large amount of organic material (from pollutants or natural sources) will allow for a greater *survival* of faecal bacteria in the marine environment. In addition to heavy metal toxicity, however, it is also apparent that substrate concentrations are generally much lower in seawater than in terrigenous environments; thus the *growth* of terrigenous bacteria would be inhibited in seawater compared with the natural marine bacteria which have adapted to very low substrate concentrations.

In summary it appears that a function of the salt content of seawater (i.e. heavy metals) together with the presence or absence of chelating agents and organic substrates are the most likely factors determining the growth and survival of terrestrial bacteria in the sea. On the other hand the division of estuarine bacteria in terms of their tolerance to sodium chloride may be a useful ecological classification with which to differentiate between autochthonous and allochthonous populations. Larsen (1962) classified micro-organims as halophobic or halophilic and described the former as organisms which grow best on a medium containing less than 2% NaCl. From an estuarine study, Seki *et al.* (1969) determined a regression line between the ratio of freshwater : saltwater colonies (y) and salinity (x). From this equation

$$y = -2{\cdot}2x + 5{\cdot}2 \tag{104}$$

it is apparent that bacteria in the estuarine environment studied by Seki *et al.* (1969) grew best in a freshwater medium ($y > 1$) when the salinity was less than 1·9% or approximately 19⁰/₀₀.

The assessment of microbial activity in polluted waters has recently been reviewed by Jannasch (1972). The author emphasizes the need to make a number of routine microbiological measurements in order to determine the biochemical activity of polluted waters, rather than simply obtaining an index of pollution, such as a coliform count. Five kinds of determinations are suggested; these are biomass determinations, respiratory activity, special metabolic activities (such as nitrification), growth rate determinations, and specificity of substrate utilization.

Estuarine areas are of particular interest to biological oceanographers, both from the point of view of their unique biological activity and because they tend to be centres of human activity which includes the development of ports. The river above the estuary may be considered as a corridor for fish migrations, ship transport, and disposal of natural and man-made products which are carried from the land to the sea. In this situation a conflict often exists between the natural ecology of the estuary and its modification due to man's activities which have generally resulted in pollution. Obvious forms of pollution, such as organic and inorganic industrial poisons, cause direct and catastrophic effects on the natural estuarine environment and remedial action is generally best handled through chemical evidence and legal procedures. Slightly more subtle, however, are the effects of non-toxic substances which may decrease light penetration in the water column, take up oxygen, or cause toxic substance to accumulate through precipitation in the sediment near the river mouth. Substances which decrease the penetration of light into the water (e.g. silt, coal dust, pulp mill effluent) can cause a decrease in the primary productivity of the water column; this in turn reduces the amount of oxygen in the water and decreases the supply of phytoplankton at the primary trophic level. Since the effect of light attenuating substances can be expressed in terms of changes in the extinction coefficient (k) of the water, a diagnosis of the overall effect of this form of pollution might be modelled through some of the expressions discussed in this text. Oxygen depletion over a period of time and distance can also be modelled using experimental data on the Biochemical Oxygen Demand (BOD) of samples of the polluting water (for BOD methodology see Amer. Pub. Health Assoc., 1965).

A property of the ecology of estuarine waters is their high productivity. In the immediate area of an estuary, where seawater and freshwater have mixed (often referred to as the river's 'plume') there is a zone of high primary production which in tropical environments, may persist throughout the year, or vary seasonally in temperate latitudes. This zone generally supports a high secondary productivity and often serves either as an area in which there is an extensive fishery, or perhaps more important, as a nursery ground for the young stages of many commercially important fish. There are a number of reasons why estuaries may be biologically productive. Firstly, there may be a natural enrichment of nutrient-poor surface sea waters by nutrients from the river; this was true for the Nile estuary, where the principal mechanism causing an enrichment of the nutrient impoverished Mediterranean waters, prior to the construction of the Aswan dam, was the nitrogen and phosphorus content of the Nile silt (Halim, 1960). Secondly, the flow of river water into the sea may cause an upwelling of nutrient-rich deep salt water which continually fertilizes the river's plume; this occurs, for example, off the Columbia and Fraser rivers. Thirdly, there may be a process of eutrophication caused by nutrients contained in the river; these are generally derived from agricultural lands and domestic sewage. Classical examples of eutrophied estuarine areas are often found in quite local estuaries (e.g. Jeffries, 1962; Barlow *et al.*, 1963) but eutrophication and upwelling may both be operative enrichment mechanisms in some of the larger estuarine areas (e.g. the New York Bight, Ketchum, 1967; the Columbia river estuary, Haertel *et al.*, 1969). In addition in some limited areas a nutrient 'trap' may develop where the organisms grown at the surface, sediment to a layer where they are carried back into the inlet by the counter current to the surface flow. This situation was first described by Redfield (1955) and examples of self-enriching processes can generally be found in fjords with shallow outer sills; in some cases the waters in the inner basins become anoxic due to the accumulation of organic materials.

The identification of nutrient polluted estuaries has been discussed by Ketchum (1967). In the simplest case it is apparent that if the nitrate or phosphate content of estuarine waters is higher than the maximum nutrient content of surrounding deep saline water, then a terrestrial source of nutrients may be expected. In the case of only two water types (i.e. freshwater and saltwater) the fraction of freshwater (F) can be obtained from the expression

$$F = \left(1 - \frac{S}{S_0}\right) \tag{105}$$

where S is the salinity of the sample and S_0 is the salinity of the source sea water. Thus the concentration of a nutrient (or some other potential pollutant) at any point in the estuary, can be determined from the salinity of a sample and a knowledge of the nutrient content of the source waters. In a system in which there are three water types Ketchum (1967) assumed that *total* phosphorus could be used as a conservative property (since loss would only be by a small amount of sedimentation of particulate phosphorus) and together with the salinity, the fraction of any three water types could be found from the salinity and total phosphorus content of a sample. Thus in Ketchum's example three water types were identified in the New York Bight; these were

A. Brackish river water: 30‰ S_a and total P_a 2·9 µg at/l
B. Surface coastal water: 30·95‰ S_b and total P_b 0·5 µg at/l
C. Deep ocean water: 34‰ S_c and total P_c 1·25 µg at/l

Then the equations used to determine the fraction of the three water types are

$$S_x = AS_a + BS_b + CS_c \tag{106}$$

$$P_x = AP_a + BP_b + CP_c \tag{107}$$

$$A + B + C = 1. \tag{108}$$

Where A, B and C are the volume fractions of each water type and S and P are the salinities and total phosphorus contents of the three water types and the unknown sample, x.

This rather simplified diagnosis of water types can be used in conjunction with other environmental data to analyse changes in the estuarine environment. For example the maximum level of primary productivity is often found at some distance from a river mouth and is not necessarily associated with the maximum availability of nutrients. This is caused by a number of factors including the increased availability of light due to sedimentation of silt, decreased mixing processes with distance, and a time factor which allows the seed population of phytoplankton to increase exponentially as the waters move away from the river mouth; this situation has been documented as occurring under natural conditions in the Fraser river plume (Parsons *et al.* 1967) and is predictable from a simulated model of a sewage outfall entering the sea (Walsh, 1972) and from actual observations (Caperon *et al.* 1971). In the latter reference the authors also use a model (similar to Walsh's—1972) to predict the course of future and past eutrophication of a tropical estuary. The model employs a 4 component food chain in which each component is related to the rest by a hyperbolic function [eqn. (68)]; feedback is also introduced through nutrient regeneration. The model shows that in an oligotrophic environment, small changes in the input of nutrients are taken up rapidly by the population acting over the maximum response region of the hyperbolic function. However, the effect of a large perturbation, in terms of a massive injection of nutrients (e.g. sewage) could not be dampened out once the rate-compensating capacity of any trophic level had been exceeded. This resulted in a build-up of nutrients and the establishment of a large biomass of a new population at the primary trophic level; in other words the original ecosystem had become destabilized.

The effect of low nutrient additions on an oligotrophic environment has been considered experimentally (Parsons et al., 1972; LeBrasseur and Kennedy, 1972; Barraclough and Robinson, 1972) and theoretically (McAllister et al., 1972) in the fertilization of a 12 000 acre lake with 100 tons of nutrients. This experiment was set up to meet the requirements of hyperbolic growth responses discussed by Caperon et al. (1971). As a result it was found possible to increase the primary productivity of the lake's ecosystem without causing instability. Further there was good evidence that the increase in primary productivity was transferred up the food chain to cause an increased production at the secondary and tertiary levels.

Chemical evidence has indicated the presence of man-made pollutants in the marine food chain and the question to the biological oceanographer is how do these materials become so widely distributed and what is the consequence of their presence? At present the best documented example of a widely distributed pollutant is DDT and other chlorinated hydrocarbons. Reports on the distribution of these pesticides include measurements of their presence at great distances from the source of pollution (e.g. Antarctica—George and Frear, 1966) and in animals at the top of the food chain (e.g. sea birds—Risebrough et al., 1967). While pollutants can become slowly dispersed by ocean currents, or more rapidly by animal migrations, it is probable that the widespread dispersion of DDT is carried out in the atmosphere. On the other hand the accumulation of chlorinated hydrocarbons in tissues of higher animals must follow a marine food chain. In this case the lower trophic levels may contain infinitesimally small quantities of a chlorinated hydrocarbon; however, these materials are not decomposed by the marine food chain. They tend to be preferentially fat soluble, and as organisms are eaten the chlorinated hydrocarbons are extracted and concentrated by animals at the next trophic level. Thus the food chain acts as a continual extraction process and some of the highest organisms in the food chain (e.g. sea birds) may accumulate quite appreciable quantities of the poison. In some cases the level of chlorinated hydrocarbons may be sufficient to interfere with the metabolism of these animals, while the metabolism of organisms lower in the food chain may appear to be unaffected. The effect of DDT and other pesticides on lower organisms can be demonstrated, however, if these substances are used in rather high concentrations (e.g. 10 to 100 ppb). Wurster (1968) and Menzel et al. (1970) demonstrated photosynthetic inhibition of phytoplankton by chlorinated hydrocarbons. However, while there is no disagreement that qualitatively chlorinated hydrocarbons are toxic to the marine environment, the quantitative aspects of the problems indicate that at present DDT levels are generally well below any dangerous concentrations with possible exceptions to this to be found in some local nearshore areas and in particular among some marine birds; in the latter example it now appears well documented that egg shell thinning and reduced breeding success is associated with the presence of DDT residues in these animals (e.g. Hickey and Anderson, 1968; Wurster and Wingate, 1968).

Under some circumstances it is of interest to know the past history of a water mass. This may be traced in part from data on temperature and salinity but it is also possible to obtain information on the biological history of water from its oxygen and nutrient content (e.g. Park, 1967). The amount of oxygen which will dissolve in sea water at the sea surface is a function of sea water temperature and salinity. Any difference between the measured oxygen content of sea water and that expected from the known solubility at the temperature and salinity of the sample is called the 'apparent oxygen utilization' or AOU (Redfield, 1942). AOU may be negative (due to photosynthesis) or positive (due to mineralization). For example, if surface water is caused to sink at a convergence, oxidation of organic material will proceed at aphotic depths and the oxygen content of the water will be lowered (positive AOU). Assuming an elemental atomic ratio in plankton of $O : C : N : P \equiv 276 : 106 : 16 : 1$, the AOU can be expressed in terms of the oxidized state of these elements in a seawater sample. Thus the *measured* nitrate and phosphate ($P_{meas.}$ and $N_{meas.}$) in a seawater sample can be expressed as

$$P_{meas.} = P_p + P_{ox} \tag{109}$$

and

$$N_{meas.} = N_p + N_{ox}, \tag{110}$$

where P_p and N_p are the levels of preformed phosphate and nitrate, and P_{ox} and N_{ox} are the levels of phosphate and nitrate derived from the apparent oxygen utilization (AOU), respectively. Equations (109) and (110) can therefore be rewritten in terms of AOU, as

$$P_{meas.} = P_p + 0.0036 \text{ AOU} \tag{111}$$

and,

$$N_{meas.} = N_p + 0.058 \text{ AOU} \tag{112}$$

where all units are in numbers of atoms (e.g. µg at/l). Sugiura (1965) has suggested that the preformed nutrients are a conservative property of sea water; as such, they can be used to characterize water masses by solving eqns. (111) and (112). The use of AOU values is not without difficulties, however, since it is assumed that no oxygen is lost to the atmosphere from the surface (Stefánsson and Richards, 1964). Loss of oxygen occurs when the surface concentration is greater than saturation. Since AOU is calculated from the difference between saturation and measured concentration, difficulties arise when the water leaves the surface with an oxygen content different from the equilibrium solubility.

6.2 EXAMPLE PROBLEMS IN FISHERIES

A great deal of scientific effort has been expended on attempting to account for fluctuations in fish populations. Ahlstrom (1961) has emphasized that the two properties of a fishery which are most desirable to know are firstly the *abundance* of the species, including temporal fluctuations, and secondly the *location*, including factors which influence the aggregation or dispersion of a species. Attempts have been made to establish causal relationships between these two properties, and biotic and abiotic variables, such as winds, currents, temperature, plankton productivity, and predators. In some cases there are clearly defined relationships; for example, Uda (1961) has given the optimum temperature range for 21 different species of commercially important fish; Rae (1957) has shown a correlation between average wind strength and brood strength of haddock. In detailed studies on oceanography and the ecology of tuna, Blackburn (1965 and 1969) has concluded that there is a very general relationship between zooplankton abundance and tuna catch, but that efforts to show a definite monthly correlation between tuna and zooplankton over one degree rectangles usually failed. However, from a study of tuna distributions with temperature it was apparent that the tuna would not generally enter water colder than 20°C. Since high concentrations of food organisms often occured in areas where the temperature was below 20°C, a direct relationship between tuna and zooplankton abundance was not apparent when data were collected from purely geographical coordinates. When temperature was taken into account it was concluded that the 20° isotherm determined the overall range of the tuna, while zooplankton concentration determined the distribution of tuna within the animal's temperature range.

In general, however, predictions of abundance or location of fish based on single or multiple component correlations have not been markedly successful. This conclusion has been reached independently by a number of authors studying very different associations in the oceans. Thus Ahlstrom (1961) concluded that there was no relationship between year class strength of sardines and high plankton productivity. Lillelund (1965) concluded from a review of the effects of abiotic factors on young stages of marine fish that abiotic factors had only an indirect effect on survival, and that the overall effects were complex and were more probably associated with biotic factors. Steele (1964) concluded that the prediction of fisheries could not be based on levels of primary productivity and assumptions regarding ecological efficiencies, but that more detailed knowledge of food-web structure was required.

While relationships between adult stocks and biological oceanographic data are tenuous, several authors have recognized the importance of survival rates during the early stages of the life history of a fish (e.g. Dementjeva, 1965; Gulland 1965; Shelbourne, 1957). Thus while most marine teleosts produce large numbers of eggs, only a few animals survive to become adults. In the case of cod, the numbers of eggs laid in the lifetime of an adult amounts to several millions, and yet on the average only two cod survive to become adults. Gulland (1965) has estimated that the annual mortality of commercially exploited adult cod, does not exceed 30 to 75% but that in the first few months of a cod's life, mortality must be as high as 99·9999%. The important point in these approximations is not in fact the mortality of the early life stages, but the 0·0001% survival; obviously conditions which might lead to a change in the survival rate from one tenthousandths of a percent to two or three tenthousandths of a percent are more easily conceived as factors in determining a large year class, than other factors, which would have to affect the survival of the more mature fish stages by several hundred percent in order to account for the wide fluctuations in fish abundance which are observed in nature. In as much as the early life stages of many fish are planktonic (e.g. herring, plaice, cod, tuna), their survival in the plankton community should be an integral part of biological oceanographic studies.

From detailed observations on the survival of plaice larvae in the North Sea, Shelbourne (1957) has described larvae in a deteriorating condition during the month of January when suitable planktonic food items were scarce. Deterioration of larvae started as the yolk reserves became exhausted; similar observations conducted during March showed the presence of healthy larvae which were feeding on plant and animal plankton. The importance of specific food items in the diet of plaice larvae was also observed; during the earliest feeding stage the larvae fed off some large diatoms (e.g. *Biddulphia* and *Coscinodiscus*) but an abrupt change to a zooplankton diet (primarily *Oikopleura*) occured very soon after feeding started.

The feeding and survival of herring larvae has been studied extensively both in the field (e.g. Blaxter, 1963, Lisivnenko, 1961) and under laboratory conditions (e.g. Rosenthal and Hempel, 1970). Blaxter's (1963) detailed observations have shown that there is some obvious selectivity for prey items depending on size, while visual sighting and attacks on individual

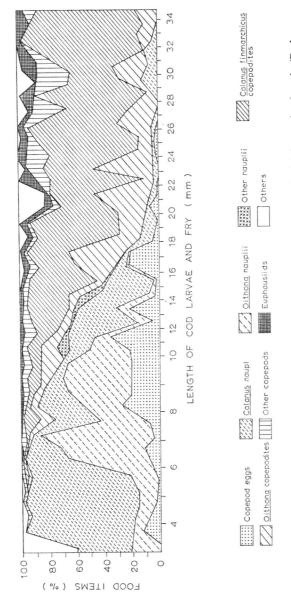

FIG. 56. Changes in the composition of the food of larvae and pelagic fry of cod with increasing length. (Redrawn from Sysoeva and Degtereva, 1965.)

prey items depended to some extent on the amount of light available. Nutritional differences between prey items consumed were also demonstrated to have an effect on the growth and survival of the larvae. From field studies conducted by a number of authors, Blaxter (1963) concluded that larvae were most abundant when food items were present at a concentration of *ca* 30 organisms per litre. This figure is similar to the concentration of prey items determined experimentally by Rosenthal and Hempel (1970), if it is assumed that only a certain fraction of the prey items seen by a larvae are effectively captured. Lisivnenko (1961) studied the abundance of herring larvae on the concentration of food organisms over a period of 5 years in the Gulf of Riga; her results show a strong correlation between an approximate 5-fold increase in larval abundance and an increase in food items from *ca* 5 to 20 organisms/litre.

Sysoeva and Degtereva (1965) showed that the main food item of the Arctic–Norwegian cod larvae and fry was the copepod, *Calanus finmarchicus*. Both larvae and fry fed off *C. finmarchicus* during different stages of the copepods life cycle and this is illustrated in Fig. 56 together with the secondary importance of *Oithona* nauplii and some other food organisms. The absolute concentration of food organisms required by cod fry for successful feeding was expressed as the number of organisms per m² in a 50 m water column. Concentrations of > 18 000 *C. finmarchicus*/m² provided a sufficient food supply while the intensity of feeding decreased over the range from 18 000 down to 5000 organisms/m².

From these examples of larval fish survival and the concentration of food organisms, it is apparent that there is an interdependent relationship which may be generally expressed as follows: above a certain concentration of organisms, survival of larvae will be high and largely independent of further increases in the concentration of food items. Over a range of concentrations below this maximum threshold for survival, larval feeding success will depend on the concentration of food items; however, at some lower threshold concentration of food items, the absorption of yolk sacs will terminate the survival of most larvae since they must have a certain minimum amount of food in order to exist as effective predators on the plankton. At this point in the predator/prey relationship, the energy that is being expended in obtaining food is equal or less than the energy obtained from the food; this effectively

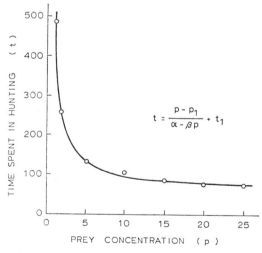

$$t = \frac{p - p_1}{\alpha - \beta p} + t_1$$

FIG. 57. Dependence of the time spent in hunting by a larval fish on the concentration of food organisms (redrawn from Ivlev, 1944).

reduces the benefit from the hunt to zero and the population of larvae dies. Ivlev (1944) expressed these observations as an equation which may be of some value in examining larval survival and food item concentration. This relationship is represented in Fig. 57 as the time spent in hunting by a larval fish predator (t) plotted against the corresponding prey density (p). In the equation for the curve, α and β are constants and (p_1) and (t_1) are arbitrarily chosen points on the curve. The curve takes the form of a hyperbola and the asymptotes on the axes have real biological significance. The asymptote parallel to the prey axis will define the predator's limiting rate of food consumption (i.e. rate of food consumption which is independent of the concentration of food items); the asymptote parallel to the t-axis represents the minimum concentration of prey at which the predator can obtain a ration and survive.

The type of relationship given above appears to be particularly important as an indication of the processes involved in the dependence of fish larvae on their food supply. Among some species of fish, however, there is essentially no pelagic larval stage and the animals emerge into the pelagic environment as juvenile fish (*ca* 35 mm length) with a much greater ability to search for prey than is found among pelagic fish larvae. Perhaps the most important example of this group of fish are the salmonids. Processes governing the early pelagic life of young salmonids with respect to their food supply in the plankton community are different in some respects from those governing larval fish survival.

The ability of juvenile fish to catch prey will depend on individual species and environments, but an indication of some of the mechanisms involved have been given in studies by a number of authors (e.g. Paloheimo and Dickie, 1965, and 1966a; LeBrasseur, 1969; Leong and O'Connell, 1969; Parsons and LeBrasseur, 1970; Parker, 1971).

From extensive studies of fish feeding data reported by various authors Paloheimo and Dickie (1966b) concluded that growth efficiency (K_1) was related to the amount of food eaten by an animal (R) such that,

$$\ln K_1 = -a - bR. \tag{113}$$

When $\ln K_1$ is plotted against R for different rations, different values of a and b are obtained. The relationship may not apply to hand fed laboratory fish cultures but there is considerable justification for its use under field conditions. The importance of eqn. (113) from the point of view of the production of food items is that constants 'a' and 'b' must be related in some way to the metabolic cost of behaviour patterns associated with particular types of prey. As an example of this effect the authors suggested that the size of prey items might be one factor in determining differences between growth efficiency and ration (i.e. in the differences in 'a' and 'b'). Differences in the ability of young salmonids to graze off different-sized prey has been demonstrated in short term experiments, such as those reported by Parsons and LeBrasseur (1970). In the latter experiments the same biomass of three different sized zooplankton (2 species of copepods and a euphausiid) were fed to young salmon over a concentration range from 0·5 to 100 g/m³ wet weight of zooplankton. The results showed that the salmon were able to eat more of the medium-size copepods per unit time than of the large euphausiids or the very small copepods. From this observation it may be concluded that not only is the absolute abundance of prey items important to the growth of juvenile fish, but also there are marked differences in the efficiency of fish growth depending on the type of prey as indicated by eqn. (113). With some fish species the method of capturing prey may change and this will also alter the rate of feeding at similar prey densities. This is illustrated in Fig. 58 from Leong and O'Connell (1969). The experimental data show that the Northern Anchovy obtained small *Artemia* nauplii by filter feeding at a slower rate

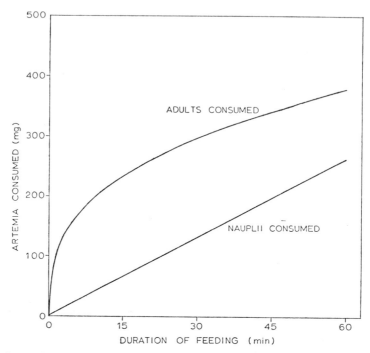

FIG. 58. Comparison of particulate and filter feeding rates for a 4 g anchovy. Curve showing 'adults consumed' was determined for *Artemia* concentrations of 2 to 50 mg adults/litre; curve showing 'nauplii consumed' was determined for *Artemia* concentrations of 4·4 mg nauplii/litre. Figure redrawn from Leong and O'Connell (1969).

than they obtain *Artemia* adults by raptorial feeding. Furthermore the shapes of the feeding curves are different and show that the greatest difference occurs at very low concentrations of prey.

From studies carried out by Parker (1971) it is also apparent that the rate at which a juvenile fish grows is an integral part of its chances for survival. Thus in the case of young pink salmon, Parker (1971) was able to show that the true growth rate of a pink salmon

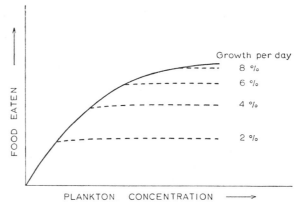

FIG. 59. Schematic relationship between plankton concentration, food eaten and the daily growth rate of juvenile fish.

population was 1·4%/day and that this growth rate was greater than that of the pink salmon's predator (2 year old coho salmon) which grew at ·7% per day. Thus the survival of the juvenile pink salmon was assured as they "outgrew" being suitable prey items for coho salmon. This type of mechanism is illustrated in Fig. 59 where it is seen that depending on the amount of zooplankton present young pink salmon may grow at different rates. Those that grow faster than the growth rate of the predator will eventually escape attacks as suitable prey items.

The associations described in this section between the plankton community and fisheries serve as illustrations of how biological oceanographic data may be utilized in understanding mechanisms contributing to the survival of fish. More complex treatments of this subject, particularly those which include the physiological requirements of a fish while hunting for its prey, have been discussed by a number of authors. In particular a recent presentation by Kerr (1971) gives an indication of how growth efficiency must include a consideration of the animal's metabolic requirements for hunting, basal metabolism and the internal cost of food utilization. These factors must then be related to the size and concentration of prey available, and the swimming speed of the predator.

REFERENCES

ABDULLAH, M. I., L. G. ROYLE and A. W. MORRIS. 1972. Heavy metal concentration in coastal waters. *Nature*, **235**, 158–160.

ACKMAN, R. G., C. S. TOCHER and J. MCLACHLAN. 1968. Marine phytoplankter fatty acids. *J. Fish. Res. Bd. Canada*, **25**, 1603–1620.

ACKMAN, R. G., C. A. EATON, J. C. SIPOS, S. N. HOOPER and J. D. CASTELL. 1970. Lipids and fatty acids of two species of North Atlantic krill *Meganyctiphanes norvegica* and *Thysanoessa inermis* and their role in the aquatic food web. *J. Fish. Res. Bd. Canada*, **27**, 513–533.

ADAMS, J. A. and J. H. STEELE. 1966. Shipboard experiments on the feeding of *Calanus finmarchicus* (Gunnerus). In *Some Contemporary Studies in Marine Science*. Ed. H. Barnes, Allen and Unwin Ltd. London, 19–35.

AHLSTROM, E. H. 1961. Fisheries oceanography. *Calif. Coop. Oceanic Fish Invest.*, **8**, 71–72.

ALLEN, M. B. 1956. Excretion of organic compounds by *Chlamydomonas*. *Arch. Mikrobiol.*, **24**, 163–168.

ALLEN, M. B. 1963. Nitrogen fixing organisms in the sea. In *Symposium on Marine Microbiology*. Ed. C. H. Oppenheimer, C. C. Thomas, Springfield, Illinois, 85–105.

ALLEN, M. B., T. W. GOODWIN and S. PHAGPOLNGARM. 1960. Carotenoid distribution in certain naturally occurring algae and in some artificially induced mutants of *Chlorella pyrenoidosa*. *J. gen. Microbiol.*, **23**, 93–103.

ALLEN, M. B., L. FRIES, T. W. GOODWIN and D. M. THOMAS. 1964. The carotenoids of algae. Pigments from some cryptomonads, a heterokont and some Rhodophyceae. *J. gen. Microbiol.* **34**, 259–267.

ALVERSON, D. L., A. R. LONGHURST and J. A. GULLAND. 1970. How much food from the sea? *Science*, **168**, 503–505.

Amer. Publ. Health Assn., Amer. Water Works Assn., Water Pollution Contr. Fed. 1965. Standard Methods for the Examination of Water and Waste Water including Bottom Sediments and Sludges. 12th ed. Amer. Publ. Health Assn. Inc. New York, pp. 769.

ANDERSON, G. C. 1964. The seasonal and geographic distribution of primary productivity off the Washington and Oregon Coasts. *Limnol. Oceanogr.*, **9**, 284–302.

ANDERSON, G. C. 1965. Fractionation of phytoplankton communities off the Washington and Oregon Coasts. *Limnol. Oceanogr.*, **10**, 477–480.

ANDERSON, G. C. 1969. Subsurface chlorophyll maximum in the northeast Pacific Ocean. *Limnol. Oceanogr.*, **14**, 386–391.

ANDERSON, G. C. and R. P. ZEUTSCHEL. 1970. Release of dissolved organic matter by marine phytoplankton in coastal and offshore areas of the northeast Pacific Ocean. *Limnol. Oceanogr.*, **15**, 402–407.

ANDREWS, P., and P. J. LEB. WILLIAMS. 1971. Heterotrophic utilization of dissolved organic compounds in the sea. III. Measurement of the oxidation rates and concentrations of glucose and amino acids in sea water. *J. mar. biol. Ass. U.K.*, **51**, 111–125.

ANRAKU, M., and M. OMORI. 1963. Preliminary survey of the relationship between the feeding habit and the structure of the mouthparts of marine copepods. *Limnol. Oceanogr.*, **8**, 116–126.

ANTIA, N. J., C. D. MCALLISTER, T. R. PARSONS, K. STEPHENS and J. D. H. STRICKLAND. 1963. Further measurements of primary production using a large-volume plastic sphere. *Limnol. Oceanogr.*, **8**, 166–183.

ANTIA, N. J., and E. BILINSKI. 1967. A bacterial toxin type of phospholipase (Lecethinase C) in a marine phytoplanktonic chrysomonad. *J. Fish. Res. Bd. Canada*, **24**, 201–204.

ANTIA, N. J., J. Y. CHENG and F. J. R. TAYLOR. 1969. The heterotrophic growth of a marine photosynthetic cryptomonad (*Chroomonas salina*). *Proc. Intl. Seaweed Symp.*, **6**, 17–29.

APSTEIN, C. 1910. Hat ein Organismus in der Tiefe gelebt, in der er gefischt ist? *Int. Rev. ges. Hydrobiol. Hydrogr.*, **3**, 17–33.

ARMSTRONG, F. A. J. and W. R. G. ATKINS. 1950. The suspended matter of sea water. *J. mar. biol. Ass. U.K.*, **29**, 139–143.

ARMSTRONG, F. A. J., and E. C. LaFOND. 1966. Chemical nutrient concentrations and their relationship to internal waves and turbidity off southern California. *Limnol. Oceanogr.*, **11**, 538–547.

ARMSTRONG, F. A. J., C. R. STEARNS and J. D. H. STRICKLAND. 1967. The measurement of upwelling and subsequent biological processes by means of the Technicon Autoanalyser ® and associated equipment. *Deep-Sea Res.*, **14**, 381–389.

ARUGA, Y. 1965a. Ecological studies of photosynthesis and matter production of phytoplankton. I. Seasonal changes in photosynthesis of natural phytoplankton. *Bot. Mag., Tokyo*, **78**, 280–288.

ARUGA, Y. 1965b. Ecological studies of photosynthesis and matter production of phytoplankton. II. Photosynthesis of algae in relation to light intensity and temperature. *Bot. Mag., Tokyo*, **78**, 360–365.

ARUGA, Y. 1966. Ecological studies of photosynthesis and matter production of phytoplankton. III. Relationship between chlorophyll amount in water and primary productivity. *Bot. Mag., Tokyo*, **79**, 20–27.

ARUGA, Y., and S. ICHIMURA, 1968. Characteristics of photosynthesis of phytoplankton and primary production in the Kuroshio. Bull. Misaki Marine Biol. Inst. Kyoto Univ., No. 12 (Proceedings of the U.S.–Japan Seminar on Marine Microbiology. August 1966 in Tokyo), 3–20.

ARUGA, Y., Y. YOKOHAMA and M. NAKANISHI. 1968. Primary productivity studies in February–March in the Northwestern Pacific off Japan. *J. Oceanogr. Soc. Japan*, **24**, 275–280.

BAAS BECKING, L. G. M. and E. J. F. WOOD. 1955. Biological processes in the estuarine environment. I and II. Ecology of the sulfur cycle. *Koninkl. Ned. Akad. Wetenschap. Proc.*, **B58**, 160–181.

BAINBRIDGE, R. 1957. The size, shape and density of marine phytoplankton concentrations. *Biol. Rev.*, **32**, 91–115.

BANSE, K. 1964. On the vertical distribution of zooplankton in the sea. *Prog. Oceanogr.*, **2**, 56–125.

BARBER, R. T. 1966. Interaction of bubbles and bacteria in the formation of organic aggregates in sea water. *Nature*, **211**, 257–258.

BARBER, R. T., R. C. DUGDALE, J. J. MacISAAC and R. L. SMITH. 1971. Variations in phytoplankton growth associated with the source and conditioning of upwelling water. *Inv. Pesq.*, **35**, 171–193.

BARLOW, J. P., C. J. LORENZEN and R. T. MYREN. 1963. Eutrophication of a tidal estuary. *Limnol. Oceanogr.*, **8**, 251–262.

BARNES, H. 1949. On the volume measurement of water filtered by a plankton pump, with an observation on the distribution of plankton animals. *J. mar. biol. Ass. U.K.*, **28**, 651–662.

BARNES, H. 1952. The use of transformations in marine biological statistics. *J. Cons. Int. Explor. Mer*, **18**, 61–71.

BARNES, H. 1956. *Balanus balanoides* (L.) in the Firth of Clyde: the development and annual variation of the larval population, and the causative factors. *J. Anim. Ecol.*, **25**, 72–84.

BARNES, H. 1959. *Oceanography and Marine Biology: A Book of Techniques*. Hafner (New York), pp. 218.

BARNES, H. and S. M. MARSHALL. 1951. On the variability of replicate plankton samples and some application of "contagious" series to the statistical distribution of catches over restricted periods. *J. mar. biol. Ass. U.K.*, **30**, 233–263.

BARRACLOUGH, W. E., R. J. LeBRASSEUR and O. D. KENNEDY. 1969. Shallow scattering layer in the subarctic Pacific Ocean: detection by high-frequency echo sounder. *Science*, **166**, 611–613.

BARRACLOUGH, W. E., and D. ROBINSON. 1972. The fertilization of Great Central Lake. III. Effect on juvenile sockeye salmon. *Fish. Bull.*, **70**, 37–48.

BARY, B. McK. 1959. Species of zooplankton as a means of identifying different surface waters and demonstrating their movements and mixing. *Pac. Sci.*, **13**, 14–34.

BARY, B. McK. 1963. Distributions of Atlantic pelagic organisms in relation to surface water bodies. In *Marine Distributions*, Ed. M. J. Dunbar, Royal Soc. Canada, Sp. Publ., No. 5: 51–67.

BARY, B. McK. 1967. Diel vertical migrations of underwater scattering, mostly in Saanich Inlet, British Columbia. *Deep-Sea Res.*, **14**, 35–50.

BAYLOR, E. R. and W. H. SUTCLIFFE. 1963. Dissolved organic matter in sea water as a source of particulate food. *Limnol. Oceanogr.*, **8**, 369–371.

BEATTIE, A., E. L. HIRST and E. PERCIVAL. 1961. Studies on the metabolism of the Chrysophyceae. Comparative structural investigations on leucosin (Chrysolaminarin) separated from diatoms and laminarin from the brown algae. *Biochem. J.*, **79**, 531–537.

BEERS, J. R. 1966. Studies on the chemical composition of the major zooplankton groups in the Sargasso Sea off Bermuda. *Limnol. Oceanogr.*, **11**, 520–528.

BEERS, J. R., M. R. STEVENSON, R. W. EPPLEY and E. R. BROOKS. 1971. Plankton populations and upwelling off the coast of Peru, June 1969. *Fish. Bull.*, **69**, 859–876.

BEKLEMISHEV, K. V. 1954. The feeding of some common plankton copepods in far eastern seas. *Zool. J., Inst. Oceanol. Acad. Sci. USSR*, **33**, 1210–1230.

BELL, R. K., and F. J. WARD. 1970. Incorporation of organic carbon by *Daphnia pulex*. *Limnol. Oceanogr.*, **15**, 713–726.

BERNARD, F. 1963. Vitesses de chute en mer des amas palmelloides de *Cyclococcolithus*. Ses conséquences pour le cycle vital des mers chaudes. *Pelagos*, **1**, 5–34.

BLACKBURN, M. 1965. Oceanography and the ecology of tunas. *Oceanogr. Mar. Biol. Ann. Rev.*, **3**, 299–322.

BLACKBURN, M. 1969. Conditions related to upwelling which determine distribution of tropical tunas off western Baja California. *Fish. Bull.*, **68**, 147–176.

BLACKBURN, M., R. M. LAURS, R. W. OWEN and B. ZEITSCHEL. 1970. Seasonal and areal changes in standing stocks of phytoplankton, zooplankton and micronekton in the eastern tropical Pacific. *Mar. Biol.*, **7**, 14–31.

BLACKMAN, F. F. 1905. Optima and limiting factors. *Ann. Bot.*, **19**, 281–295.

BLAXTER, J. H. S. 1963. The feeding of herring larvae and their ecology in relation to feeding. *Calif. Coop. Oceanic Fish Investig. Rep.*, **10**, 79–88.

BLUMER, M., M. M. MULLIN and R. R. L. GUILLARD. 1970. A polyunsaturated hydrocarbon (3, 6, 9, 12, 15, 18-heneicosahexaene) in the marine food web. *Mar. Biol.*, **6**, 226–235.

BLUMER, M., R. R. L. GUILLARD and T. CHASE. 1971. Hydrocarbons of marine phytoplankton. *Mar. Biol.*, **8**, 183–189.

BOYSEN JENSEN, P. 1932. Die Stoffproduktion der Pflanzen. *Publ. Gustav Fischer (Jena)*, pp. 108.

BRAARUD, T. 1963. Reproduction in the marine coccolithophorid *Coccolithus huxleyi* in culture. *Pubbl. staz. zool. Napoli*, **33**, 110–116.

BRAARUD, T., and B. FØYN. 1931. Beiträge zur Kenntnis des Stoffwechsels in Meer. *Avh. norske Vidensk Akad. Oslo*, **14**, pp. 24.

BROCKSEN, R. W., G. E. DAVIS, and C. E. WARREN. 1970. Analysis of trophic processes on the basis of density-dependent functions. In *Marine Food Chains*, Ed. J. H. Steele. Oliver and Boyd Edinburgh, 468–498.

BROOKS, J. L., and S. I. DODSON. 1965. Predation, body size, and composition of plankton. *Science*, **150**, 28–35.

BROWN, T. E., and F. L. RICHARDSON. 1968. The effect of growth environment on the physiology of algae: light intensity. *J. Phycol.*, **4**, 38–54.

BUNKER, H. J. 1936. A review of the physiology and biochemistry of the sulfur bacteria. *Dept. Sci. Ind. Res., Chem. Res. Spec. Rpt.*, **3**.

BUNT, J. S. 1964a. Primary productivity under sea ice in Antarctic waters. 1. Concentrations and photosynthetic activities of microalgae in the waters of McMurdo Sound, Antarctica. *Antarctic Res. Ser.*, **1**, 13–26.

BUNT, J. S. 1964b. Primary productivity under sea ice in Antarctic waters. 2. Influence of light and other factors on photosynthetic activities of Antarctic marine microalgae. *Antarctic Res. Ser.*, **1**, 27–31.

BUNT, J. S. 1965. Measurements of photosynthesis and respiration in a marine diatom with the mass spectrometer and with carbon-14. *Nature*, **207**, 1373–1375.

BUNT, J. S., and C. C. LEE. 1970. Seasonal primary production in Antarctic sea ice at McMurdo Sound in 1967. *J. Mar. Res.*, **28**, 304–320.

BURKE, V., and L. A. BAIRD. 1931. Fate of fresh water bacteria in the sea. *J. Bacteriol.*, **21**, 287–298.

BURKHOLDER, P. R., and L. M. BURKHOLDER. 1956. Vitamin B_{12} in suspended solids and marsh muds collected along the coast of Georgia. *Limnol. Oceanogr.*, **1**, 202–208.

BURKHOLDER, P. R., L. M. BURKHOLDER and L. P. ALMODÓVAR. 1960. Antibiotic activity of some marine algae of Puerto Rico. *Botanica Mar.*, **2**, 149–156.

BURKHOLDER, P. R., and E. F. MANDELLI. 1965. Productivity of microalgae in Antarctic sea ice. *Science*, **149**, 872–874.

BUTLER, E. I., E. D. S. CORNER and S. M. MARSHALL. 1970. On the nutrition and metabolism of zooplankton. VII. Seasonal survey on nitrogen and phosphorus excretion by *Calanus* in the Clyde sea area. *J. mar. biol. Ass. U.K.*, **50**, 525–560.

CALVIN, M., and J. A. BAASHAM. 1962. *The photosynthesis of carbon compounds*. Benjamin, New York, pp. 127.

CAPERON, J. 1967. Population growth in micro-organisms limited by food supply. *Ecology*, **48**, 715–722.

CAPERON, J. 1968. Population growth response of *Isochrysis galbana* to nitrate variation at limiting concentrations. *Ecology*, **49**, 866–872.

CAPERON, J., S. A. CATTELL and G. KRASNICK. 1971. Phytoplankton kinetics in a subtropical estuary: eutrophication. *Limnol. Oceanogr.*, **16**, 599–607.

CARLUCCI, A. F., and P. M. WILLIAMS. 1965. Concentration of bacteria from seawater by bubble scavenging. *J. Cons. Int. Explor. Mer*, **30**, 28–33.

CARLUCCI, A. F., and P. M. McNALLY. 1969. Nitrification by marine bacteria in low concentrations of substrate and oxygen. *Limnol. Oceanogr.*, **14**, 736–739.

CARLUCCI, A. F., and P. M. BOWES. 1970. Production of vitamin B_{12}, thiamine, and biotin by phytoplankton. *J. Phycol.*, **6**, 351–357.

CARPENTER, E. J., and R. R. L. GUILLARD. 1971. Intraspecific differences in nitrate half-saturation constants for three species of marine phytoplankton. *Ecology*, **52**, 183–185.

CASSIE, R. M. 1959. Micro-distribution of plankton. *New Zealand J. Sci.*, **2**, 398–409.

CASSIE, R. M. 1962a. Frequency distribution models in the ecology of plankton and other organisms. *J. Anim. Ecol.*, **31**, 65–92.

CASSIE, R. M. 1962b. Microdistribution and other error components of C^{14} primary production estimates. *Limnol. Oceanogr.*, **7**, 121–130.

CASSIE, R. M. 1963. Multivariate analysis in the interpretation of numerical plankton data. *New Zealand J. Sci.*, **6**, 36–59.

CASSIE, R. M. 1968. Sample design. Zooplankton Sampling. *Mon. Oceanogr. Method. Unesco (Paris)*, **2**, 105–121.

CHAU, Y. K., L. CHUECAS and J. P. RILEY. 1967. The component combined amino acids of some marine phytoplankton species. *J. mar. biol. Ass. U.K.*, **47**, 543–554.

CHAVE, K. E. 1965. Carbonates: association with organic matter in surface seawater. *Science*, **148**, 1723–1724.

CHAVE, K. E. 1970. Carbonate-organic interactions in sea water. *Symp. Organic Matter in Natural Waters.* Ed. D. W. Hood, University of Alaska, 373–385.

CHINDONOVA, Y. G. 1959. The nutrition of certain groups of abyssal macroplankton in the north-western area of the Pacific Ocean. *Trudy Institute Okeanologii*, **30**, 166–189. (Nat. Inst. Oceanogr. Translation No. 131, Wormley, U.K.).

CHISLENKO, L. L. 1968. Nomographs for determination of weights of aquatic organisms by size and body form (marine mesobenthos and plankton). Acad. Sci. USSR. Zool. Inst. "Science" Publishing House, Leningrad. (Translation by J. Marliave, Univ. British Columbia.)

CHUECAS, L., and J. P. RILEY. 1969. Component fatty acids of the total lipids of some marine phytoplankton. *J. mar. biol. Ass. U.K.*, **49**, 97–116.

COLEBROOK, J. M., R. S. GLOVER and G. A. ROBINSON. 1961. Continuous plankton records: contributions towards a plankton atlas of the north-eastern Atlantic and the North Sea. *Bull. Mar. Ecol.*, **5**, 67–80.

COMITA, G. W. 1968. Oxygen consumption in *Diaptomus*. *Limnol. Oceanogr.*, **13**, 51–57.

CONOVER, R. J. 1966a. Assimilation of organic matter by zooplankton. *Limnol. Oceanogr.*, **11**, 338–345.

CONOVER, R. J. 1966b. Factors affecting the assimilation of organic matter by zooplankton and the question of superfluous feeding. *Limnol. Oceanogr.*, **11**, 346–354.

CONOVER, R. J. 1966c. Feeding on large particles by *Calanus hyperboreus* (Kröyer). In *Some Contemporary Studies in Marine Science*. Ed. H. Barnes, Allen and Unwin, London, 187–194.

CONOVER, R. J. 1968. Zooplankton—life in a nutritionally dilute environment. *Am. Zoologist*, **8**, 107–118.

COOMBS, J., W. M. DARLEY, O. HOLM-HANSEN and B. E. VOLCANI. 1967a. Studies on the biochemistry and fine structure of silica shell formation in diatoms. Chemical composition of *Navicula pelliculsa* in silicon-starvation synchrony. *Plant Physiol.*, **42**, 1601–1606.

COOMBS, J., C. SPANIS and B. E. VOLCANI. 1967b. Studies on the biochemistry and fine structure of silica shell formation in diatoms. Photosynthesis and respiration in silicon-starvation synchrony of *Navicula pelliculosa*. *Plant Physiol.*, **42**, 1607–1611.

COOMBS, J., P. J. HALICKI, O. HOLM-HANSEN and B. E. VOLCANI. 1967c. Studies on the biochemistry and fine structure of silica shell formation in diatoms. II. Changes in the concentration of neucloside triphosphates in silicon starvation synchrony of *Navicula pelliculosa*. (Bréb.) *Hilse. Exptl. Cell Res.*, **47**, 315–328.

COOPER, L. H. N. 1937. On the ratio of nitrogen to phosphorus in the sea. *J. mar. biol. Ass. U.K.*, **22**, 177–182.

CORNER, E. D. S. 1961. On the nutrition and metabolism of zooplankton. I. Preliminary observations on the feeding of the marine copepod, *Calanus helgolandicus* (Claus). *J. mar. biol. Ass. U.K.*, **41**, 5–16.

CORNER, E. D. S., and B. S. NEWELL. 1967. On the nutrition and metabolism of zooplankton. IV. The forms of nitrogen excreted by *Calanus*. *J. mar. biol. Ass. U.K.*, **47**, 113–120.

CORNER, E. D. S., C. B. COWEY and S. M. MARSHALL. 1967. On the nutrition and metabolism of zooplankton. V. Feeding efficiency of *Calanus finmarchicus*. *J. mar. biol. Ass. U.K.*, **47**, 259–270.

COWEY, C. B., and E. D. S. CORNER. 1963. On the nutrition and metabolism of zooplankton. II. The relationship between the marine copepod *Calanus helgolandicus* and particulate material in Plymouth sea water, in terms of amino acid composition. *J. mar. biol. Ass. U.K.*, **43**, 495–511.

COWEY, C. B., and E. D. S. CORNER. 1966. The amino-acid composition of certain unicellular algae, and of the faecal pellets produced by *Calanus finmarchicus* when feeding on them. In *Some Contemporary Studies in Marine Science*. Ed. H. Barnes, George Allen and Unwin, London, 225–231.

CULKIN, F., and R. J. MORRIS. 1970. The fatty acid composition of two marine filter-feeders in relation to a phytoplankton diet. *Deep-Sea Res.*, **17**, 861–865.

CUPP, E. E. 1943. *Marine plankton diatoms of the west coast of North America*. University of California Press, Berkeley, pp. 237.

CURL, H. 1962. Standing crops of carbon, nitrogen, and phosphorus and transfer between trophic levels, in continental shelf waters south of New York. *Rapp. Proc.-Verb. Cons. int. Explor. Mer*, **153**, 183–189.

CURRIE, R. I. 1962. Pigments in zooplankton faeces. *Nature*, **193**, 956–957.

CUSHING, D. H. 1955. Production of a pelagic fishery in the sea. *Fish. Invest. London, Ser. 2*, **18**, p. 103.

CUSHING, D. H. 1962. Patchiness. *Rapp. Proc.-Verb. Cons. int. Explor. Mer*, **153**, 152–164.

CUSHING, D. H. 1964. The work of grazing in the sea. In *Grazing in Terrestrial and Marine Environments*. Ed. D. J. Crisp. Blackwell, London, 207–225.

CUSHING, D. H., and T. VUCETIC. 1963. Studies on a *Calanus* patch. III. The quantity of food eaten by *Calanus finmarchicus*. *J. mar. biol. Ass. U.K.*, **43**, 349–371.

CUSHING, D. H., and H. F. NICHOLSON. 1966. Method of estimating algal production rates at sea. *Nature*, **212**, 310–311.

CUSHMAN, J. A. 1931. The Foraminifera of the Atlantic Ocean. *Bull. U.S. Nat. Mus.*, **104**, pp. 55.

DAL PONT, G., and B. NEWELL. 1963. Suspended organic matter in the Tasman Sea. *Aust. J. Mar. Freshw. Res.*, **14**, 155–165.

DALES, R. P. 1960. On the pigments of the Chrysophyceae. *J. mar. biol. Ass. U.K.*, **39**, 693–699.

DEGENS, E. T. 1970. Molecular nature of nitrogenous compounds in sea water and recent marine sediments. In *Organic Matter in Natural Waters*. Ed. D. W. Hood. University of Alaska, pp. 77–106.

DEMENTJEVA, T. F. 1965. Changes in recruitment to the stock in relation to the environment, with reference to mathematical modelling. *ICNAF Spec. Publ.*, **6**, 381–385.

DICKIE, L. M., and K. H. MANN. (1972). *Analysis of biological production systems*, pp. 37.

DOTY, M. S. 1959. Phytoplankton photosynthetic periodicity as a function of latitude. *J. mar. biol. Ass. India*, **1**, 66–68.

DOTY, M. S., and M. OGURI. 1957. Evidence for a photosynthetic daily periodicity. *Limnol. Oceanogr.*, **2**, 37–40.

DROOP, M. R. 1968. Vitamin B_{12} and marine ecology. IV. The kinetics of uptake, growth and inhibition in *Monochrysis lutheri*. *J. mar. biol. Ass. U.K.*, **48**, 689–733.

DROOP, M. R. 1970. Vitamin B_{12} and marine ecology. V. Continuous culture as an approach to nutritional kinetics. *Helgoländer wiss. Meeresunters.*, **20**, 629–636.

DROOP, M. R., J. J. A. McLAUGHLIN, I. J. PINTER and L. PROVASOLI. 1959. Specificity of some protophytes toward vitamin B-like compounds. *Preprints Int. Oceanogr. Congr. (New York)*, 916–918.

DUFF, D. C. B., D. L. BRUCE and N. J. ANTIA. 1966. The antibacterial activity of marine planktonic algae. *Can. J. Microbiol.*, **12**, 877–884.

DUGDALE, R. C. 1967. Nutrient limitation in the sea: dynamics, identification and significance. *Limnol. Oceanogr.*, **12**, 685–695.

DUGDALE, R. C., J. J. GOERING and J. H. RYTHER. 1964. High-nitrogen fixation rates in the Sargasso Sea and the Arabian Sea. *Limnol. Oceanogr.*, **9**, 507–510.

DUGDALE, R. C., and J. J. GOERING. 1967. Uptake of new and regenerated forms of nirogen in primary productivity. *Limnol. Oceanogr.*, **12**, 196–206.

DUNBAR, M. J. 1960. The evolution of stability in marine environments. Natural selection at the level of the ecosystem. *Am. Nat.*, **94**, 129–136.

DUSSART, B. M. 1965. Les différentes catégories de planction. *Hydrobiologia*, **26**, 72–74. (with Erratum).

DUURSMA, E. K. 1960. *Dissolved organic carbon, nitrogen and phosphorus in the sea*. Ph.D. Thesis, J. B. Wolters, Groningen, pp. 147.

DUURSMA, E. K. 1965. The dissolved organic constituents of sea water. In *Chemical Oceanography*, Ed. J. P. Riley and G. Skirrow, Academic Press, New York, 433–475.

EMERSON, R., and C. M. LEWIS. 1942. The photosynthetic efficiency of phycocyanin *Chroococcus*, and the problem of carotenoid participation in photosynthesis. *J. Gen. Physiol.*, **25**, 579–595.

EPPLEY, R. W. 1972. Temperature and phytoplankton growth in the sea. *Fish. Bull.*, **70**, 1063–1085.

EPPLEY, R. W., and P. R. SLOAN. 1965. Carbon balance experiments with marine phytoplankton. *J. Fish. Res. Bd. Canada*, **22**, 1083–1097.

EPPLEY, R. W., R. W. HOLMES and J. D. H. STRICKLAND. 1967. Sinking rates of marine phytoplankton measured with a fluorometer. *J. exp. mar. biol. Ecol.*, **1**, 191–208.

EPPLEY, R. W., and J. L. COATSWORTH. 1968. Uptake of nitrate and nitrite by *Ditylum brightwellii*—kinetics and mechanisms. *J. Phycol.*, **4**, 151–156.

EPPLEY, R. W., O. HOLM-HANSEN and J. D. H. STRICKLAND. 1968. Some observations on the vertical migration of dinoflagellates. *J. Phycol.*, **4**, 333–340.

EPPLEY, R. W., and J. D. H. STRICKLAND. 1968. Kinetics of marine phytoplankton growth. *In Advances in Microbiology of the Sea.* Eds. M. R. Droop and E. J. F. WOOD, Academic Press, London, vol. 1, 23–62.

EPPLEY, R. W., J. L. COATSWORTH and L. SOLORZANO. 1969a. Studies of nitrate reductase in marine phytoplankton. *Limnol. Oceanogr.*, **14**, 194–205.

EPPLEY, R. W., J. N. ROGERS and J. J. MCCARTHY. 1969b. Half saturation constants for uptake of nitrate and ammonium by marine phytoplankton. *Limnol. Oceanogr.*, **14**, 912–920.

EPPLEY, R. W., and W. H. THOMAS. 1969. Comparison of half-saturation constants for growth and nitrate uptake of marine phytoplankton. *J. Phycol.*, **5**, 375–379.

EPPLEY, R. W., A. F. CARLUCCI, O. HOLM-HANSEN, D. KIEFER, J. J. MCCARTHY and P. M. WILLIAMS. 1972. Evidence for eutrophication in the sea near southern California coastal sewage outfalls, July 1970. *Calcofi Rep.*

FAGER, E. W., and J. A. MCGOWAN. 1963. Zooplankton species groups in the North Pacific. *Science*, **140**, 453–460.

FALLER, A. J. 1971. Oceanic turbulence and the Langmuir circulation *Ann. Rev. Ecol. Syst.*, **2**, 201–236.

FEE, E. J. 1969. A numerical model for the estimation of photosynthetic production, integrated over time and depth, in natural waters. *Limnol. Oceanogr.*, **14**, 906–911.

FEE, E. J. 1971. *A numerical model for the estimation of integral primary production and its application to Lake Michigan.* Ph.D. Thesis, University of Wisconsin, pp. 169.

FITZGERALD, G. P. 1968. Detection of limiting or surplus nitrogen in algae and aquatic weeds. *J. Phycol.*, **4**, 121–126.

FOERSTER, R. E. 1968. The sockeye salmon, *Oncorhynchus nerka. Bull. Fish. Res. Bd. Canada, Bull.* **162**, pp. 422.

FOERSTER, R. E., and W. E. RICKER. 1941. The effect of reduction of predaceous fish on survival of young sockeye salmon at Cultus Lake. *J. Fish. Res. Bd. Canada*, **5**, 315–336.

FOGG, G. E. 1952. The production of extracellular nitrogenous substances by a blue-green alga. *Proc. Roy. Soc. (London)*, B, **139**, 372–397.

FOGG, G. E. 1966. The extracellular products of algae. *Oceanogr. Mar. Biol. Ann. Rev.*, **4**, 195–212.

FOGG, G. E., C. NALEWAJKO and W. D. WATT. 1965. Extracellular products of phytoplankton photosynthesis. *Proc. Royal Soc. (London)*, **B**, **162**, 517–534.

FORD, C. W., and E. PERCIVAL. 1965. The carbohydrates of *Phaedactylum tricornutum*. Part 1. Preliminary examination of the organism and characterization of low molecular weight material and of a glucan. *J. Chem. Soc.*, (5), 7035–7042.

FOURNIER, R. O. 1966. North Atlantic deep-sea fertility. *Science*, **153**, 1250–1252.

FOX, D. L., D. M. UPDEGRAFF and G. D. NOVELLI. 1944. Carotenoid pigments in the ocean floor. *Arch. Biochem.*, **5**, 1–23.

FRITSCH, F. E. 1956. The Structure and Reproduction of the Algae, vol. 1. Cambridge University Press, London, pp. 791.

FUJITA, Y. 1970. Photosynthesis and plant pigments. *Bull. Plankton Soc. Japan*, **17**, 20–31.

FULTON, J. 1972. Trials with an automated plankton counter. *J. Fish. Res. Bd. Canada.* **29**, 1075–1078.

GARDINER, A. C. 1933. Vertical distribution in *Calanus finmarchicus. J. mar. biol. Ass. U.K.*, **18**, 575–610.

GARRETT, W. D. 1964. *The organic chemical composition of the ocean surface.* NRL Report, 6201, 1–12.

GARRETT, W. D. 1965. Collection of slick-forming materials from the sea surface. *Limnol. Oceanogr.*, **10**, 602–605.

GARRETT, W. D. 1967. The organic chemical composition of the ocean surface. *Deep-Sea Res.*, **14**, 221–227.

GAULD, D. T. 1951. The grazing rate of planktonic copepods. *J. mar. biol. Ass. U.K.*, **29**, 695–706.

GAULD, D. T. 1966. The swimming and feeding of planktonic copepods. In *Some Contemporary Studies in Marine Science*. Ed. H. Barnes. George Allen and Unwin, London, 313–334.

GEORGE, J. L., and D. E. H. FREAR. 1966. Pesticides in the Antarctic. *J. appl. Ecol.*, **3** (suppl.), 155–167.

GIBBS, M., and J. A. SCHIFF. 1960. Chemosynthesis: the energy relations of chemoautotrophic organisms. In *Plant Physiology, vol. 1B. Photosynthesis and Chemosynthesis*, Ed. F. C. Steward, Academic Press, London, 279–319.

GOERING, J. J., R. C. DUGDALE and D. W. MENZEL. 1966. Estimates of *in situ* rates of nitrogen uptake by *Trichodesmium* sp. in the tropical Atlantic Ocean. *Limnol. Oceanogr.*, **11**, 614–620.

GOERING, J. J., D. D. WALLEN and R. M. NAUMAN. 1970. Nitrogen uptake by phytoplankton in the discontinuity layer of the eastern subtropical Pacific Ocean. *Limnol. Oceanogr.*, **15**, 789–796.

GOLDBERG, E. D. 1971. River-ocean interactions. In *Fertility of the Sea*. Ed. J. D. Costlow, Gordon and Breach, New York, vol. 1, 143–156.

GOODWIN, T. W. 1955. Carotenoids. In *Modern Methods of Plant Analysis*. Eds. K. Paech and M. V. Traceay. Springer-Verlag (Berlin), vol. 3, 272–311.

GOODWIN, T. W. 1957. The nature and distribution of carotenoids in some blue-green algae. *J. Gen. Microbiol.*, **17**, 467–473.

GORDON, D. C. 1970a. A microscopic study of organic particles in the North Atlantic Ocean. *Deep-Sea Res.*, **17**, 175–185.

GORDON, D. C. 1970b. Some studies on the distribution and composition of particulate organic carbon in the North Atlantic Ocean. *Deep-Sea Res.*, **17**, 233–243.

GORDON, L. I., P. K. PARK, S. W. HAGER and T. R. PARSONS. 1971. Carbon dioxide partial pressures in north Pacific surface waters—time variations. *J. Oceanogr. Soc. Japan*, **27**, 81–90.

GORDON, W. G., and E. O. WHITTIER. 1966. In *Fundamentals of Dairy Chemistry*. Eds. B. H. Webb and A. H. Johnson. Avi Publishing Co. Westport, Conn., pp. 60.

GRANN, H. H., and T. BRAARUD. 1935. A quantitative study of the phytoplankton in the Bay of Fundy and the Gulf of Maine including observations on hydrography, chemistry and turbidity. *J. Biol. Bd. Canada* **1**, 219–467.

GRANT, B. R. 1967. The action of light on nitrate and nitrite assimilation by the marine chlorophyte, *Dunaliella tertiolecta* (Butcher). *J. gen. Microbiol.*, **48**, 379–389.

GREGORY, R. R. F. 1971. *Biochemistry of Photosynthesis*. Wiley–Interscience, London, pp. 202.

GRENZE, V. N., and E. P. BALDINA. 1964. *Trudy Sevastopol biol. Sta. 18*, Fish, Res. Bd. Canada Translation 893 (Population dynamics and the annual production of *Acartia clausii* Giesbr. and *Centropages kroyeri* in the neritic zone of the Black Sea).

GRIFFIN, J. J., H. WINDOM and E. D. GOLDBERG. 1968. The distribution of clay minerals in the world ocean. *Deep-Sea Res.*, **15**, 433–459.

GROSS, F., and E. ZEUTHEN. 1948. The buoyancy of plankton diatoms: a problem of cell physiology. *Proc Roy. Soc. (London,)* **135**, 382–389.

GUILLARD, R. R. L. 1963. Organic sources of nitrogen for marine centric diatoms. In *Symposium on Marine Microbiology*. Ed. C. H. Oppenheimer, C. C. Thomas, Springfield, Illinois, 93–104.

GULLAND, J. A. 1965. Survival of the youngest stages of fish, and its relation to year-class strength. *ICNAF Spec. Publ.*, **6**, 363–371.

GUNDERSEN, K. 1968. The formation and utilization of reducing power in aerobic chemo-autotrophic bacteria. *Zeitschrift. Allg. Mikrobiol.*, **8**, 445–457.

HAERTEL, L., C. OSTERBERG, H. CURL, JR. and P. K. PARK. 1969. Nutrient and plankton ecology of the Columbia River estuary. *Ecology*, **50**, 962–978.

HALIM, Y. 1960. Observations on the Nile bloom of phytoplankton in the Mediterranean. *J. Cons. int. explor. Mer*, **26**, 57–67.

HALLDAL, P. 1958. Pigment formation and growth in blue-green algae in crossed gradients of light intensity and temperature. *Physiol. Plant.*, **11**, 401–420.

HALTINER, G. J., and F. L. MARTIN. 1957. *Dynamical and physical meteorology*. McGraw Hill, New York, pp. 470.

HAMILTON, R. D., O. HOLM-HANSEN and J. D. H. STRICKLAND. 1968. Notes on the occurrence of living microscopic organisms in deep water. *Deep-Sea Res.*, **15**, 651–656.

HAMILTON, R. D., and J. E. PRESLAN. 1970. Observations on heterotrophic activity in the eastern tropical Pacific. *Limnol. Oceanogr.*, **15**, 395–401.

HANDA, N. 1968. Dissolved and particulate carbohydrates. *Symp. Organic Matter in Natural Waters*. Ed. D. W. Hood, University of Alaska, 129–152.

HANDA, N. 1969. Carbohydrate metabolism in the marine diatom *Skeletonema costatum*. *Mar. Biol.*, **4**, 208–214.

HANDA, N., and H. TOMINAGA. 1969. A detailed analysis of carbohydrates in marine particulate matter. *Mar. Biol.*, **2**, 228–235.

HANDA, N., and K. YANAGI. 1969. Studies on water extractable carbohydrates of the particulate matter from the northwest Pacific Ocean. *Mar. Biol.*, **4**, 197–207.

HARDER, W. 1968. Reactions of plankton organisms to water stratification. *Limnol. Oceanogr.*, **13**, 156–168.

HARDY, A. C. 1936. The continuous plankton recorder. *Discovery Rep.*, **11**, 457–510.

HARRIS, E. 1959. The nitrogen cycle in Long Island Sound. *Bull. Bingham Oceanogr.*, **17** (1), 31–65.

HARVEY, G. W. 1966. Microlayer collection from the sea surface: A new method and initial results. *Limnol. Oceanogr.*, **11**, 608–613.

HARVEY, H. W. 1957. *The Chemistry and Fertility of Sea Water.* 2nd edn. Cambridge University Press, pp. 234.

HATTORI, A. 1962a. Light-induced reduction of nitrate, nitrite and hydroxylamine in a blue-green alga, *Anabaena cylindrica. Plant and Cell Physiol.*, **3**, 355–369.

HATTORI, A. 1962b. Adaptive formation of nitrate reducing system in *Anabaena cylindrica. Plant and Cell Physiol.*, **3**, 371–377.

HATTORI, A., and J. MYERS. 1966. Reduction of nitrate and nitrite by subcellular preparations of *Anabaena cylindrica.* I. Reduction of nitrite to ammonia. *Plant Physiol.*, **41**, 1031–1036.

HATTORI, A., and E. WADA. 1971. Nitrite distribution and its regulating processes in the equatorial Pacific Ocean. *Deep-Sea Res.*, **18**, 557–568.

HAXO, F. T. 1960. The wavelength dependence of photosynthesis and the role of accessory pigments. In *Comparative Biochemistry of Photoreactive Systems.* Ed. M. B. Allen, Academic Press, New York, 339–360.

HAXO, F. T., and D. C. FORK. 1959. Photosynthetically active accessory pigments of cryptomonads. *Nature*, **184**, 1051–1052.

HEDGPETH, J. W. 1957. Classification of marine environments. *Mem. Geol. Soc. Amer.*, **67**, 17–28.

HEINRICH, A. K. 1962. The life histories of plankton animals and seasonal cycles of plankton communities in the oceans. *J. Cons. Int. Explor. Mer*, **27**, 15–24.

HELLEBUST, J. A. 1965. Excretion of some organic compounds by marine phytoplankton. *Limnol. Oceanogr.*, **10**, 192–206.

HELLEBUST, J. A. 1970. The uptake and utilisation of organic substances by marine phytoplankters. *In Organic matter in Natural Waters.* Ed. D. W. Hood. University of Alaska, 225–256.

HEMPEL, G. and H. WEIKERT. 1972. The neuston of the subtropical and boreal North-eastern Atlantic Ocean. A review. *Mar. Biol.*, **13**, 70–88.

HERSEY, J. B., and R. H. BACKUS. 1962. Sound scattering by marine organisms. In *The Sea.* Ed. M. N. Hill, Interscience New York, vol. 1, 498–539.

HEWITT, E. J. 1957. Some aspects of micronutrient element metabolism in plants. *Nature*, **180**, 1020–1022.

HICKEY, J. J., and D. W. ANDERSON. 1968. Chlorinated hydrocarbons and eggshell changes in raptorial and fish-eating b.irds. *Science*, **162**, 271–273.

HILL, R., and C. P. WHITTINGHAM. 1955. Photosynthesis. Methuen's Monographs. John Wiley, New York, pp. 165.

HINCHCLIFFE, P. R., and J. P. RILEY. 1972. The effect of diet and the component fatty acid composition of *Artemia salma. J. mar. biol. Ass. U.K.*, **52**, 203–211.

HOBSON, L. A. 1966. Some influences of the Columbia River effluent on marine phytoplankton during January, 1961. *Limnol. Oceanogr.*, **11**, 223–234.

HOBSON, L. A. 1971. Relationships between particulate organic carbon and micro-organisms in upwelling areas off Southwest Africa. *Invest. Pesq.*, **35**, 195–208.

HOBSON, L. A., and C. J. LORENZEN. 1972. Relationship of chlorophyll maxima to density structure in the Atlantic Ocean and Gulf of Mexico. *Deep-Sea Res.*, **19**, 297–306.

HOCK, C. W. 1940. Decomposition of chitin by marine bacteria. *Biol. Bull.*, **79**, 199–206.

HOFMANN, T., and H. LEES. 1952. The biochemistry of the nitrifying organisms. 2. The free-energy efficiency of *Nitrosomonas. Biochem. J.*, **52**, 140–142.

HOGETSU, K., M. SAKAMOTO and H. SUMIKAWA. 1959. On the high photosynthetic activity of *Skeletonema costatum* under the strong light intensity. *Bot. Mag., Tokyo*, **72**, 421–422.

HOLMES, R. W., and T. M. WIDRIG. 1956. The enumeration and collection of marine phytoplankton. *J. Cons. Int. Explor. Mer*, **22**, 21–32.

HOLM-HANSEN, O. 1968. Ecology, physiology, and biochemistry of blue-green algae. Ann. Rev. Microbiol., **22**, 47–70.

HOLM-HANSEN, O. 1969a. Determination of microbial biomass in ocean profiles. *Limnol. Oceanogr.*, **14**, 740–747.

HOLM-HANSEN, O. 1969b. Algae: amounts of DNA and organic carbon in single cells. *Science*, **163**, 87–88.

HOLM-HANSEN, O., and C. R. BOOTH. 1966. The measurement of adenosine triphosphate in the ocean and its ecological significance. *Limnol. Oceanogr.*, **11**, 510–519.

HOLM-HANSEN, O., J. D. H. STRICKLAND and P. M. WILLIAMS. 1966. A detailed analysis of biologically important substances in a profile off southern California. *Limnol. Oceanogr.*, **11**, 548–561.

HULBURT, E. M. 1970. Competition for nutrients by marine phytoplankton in oceanic, coastal, and estuarine regions. *Ecology*, **51**, 475–484.

HULBURT, E. M., J. H. RYTHER and R. R. L. GUILLARD. 1960. The phytoplankton of the Sargasso Sea off Bermuda. *J. Cons. int. explor. Mer*, **25**, 115–128.

HURD, L. E., M. V. MELLINGER, L. L. WOLFE, and S. J. McNAUGHTON. 1971. Stability and diversity of three trophic levels in terrestrial successional ecosystems. *Science*, **173**, 1134–1136.

HUTCHINSON, G. E. 1969. Eutrophication, past and present. In *Eutrophication: Causes, consequences, correctives*. Nat. Acad. Sci., Washington, D.C. 17–26.

ICHIMURA, S. 1956a. On the ecological meaning of transparency for the production of matter in phytoplankton community of lake. *Bot. Mag., Tokyo*, **69**, 219–226.

ICHIMURA, S. 1956b. On the standing crop and productive structure of phytoplankton community in some lakes of central Japan. *Bot. Mag., Tokyo*, **69**, 7–16.

ICHIMURA, S. 1960. Diurnal fluctuation of chlorophyll content in lake water. *Bot. Mag., Tokyo*, **73**, 217–224.

ICHIMURA, S. 1967. Environmental gradient and its relation to primary productivity in Tokyo Bay. *Records Oceanogr. Works, Japan*, **9**, 115–128.

ICHIMURA, S., and Y. SAIJO. 1958. On the application of ^{14}C-method to measuring organic matter production in the lake. *Bot. Mag., Tokyo*, **17**, 174–180.

ICHIMURA, S., Y. SAIJO and Y. ARUGA. 1962. Photosynthetic characteristics of marine phytoplankton and their ecological meaning in the chlorophyll method. *Bot. Mag., Tokyo*, **75**, 212–220.

ICHIMURA, S., and Y. ARUGA. 1964. Photosynthetic natures of natural algal communities in Japanese waters. In *Recent Researches in the Fields of Hydrosphere, Atmosphere and Nuclear Geochemistry*. Eds. Y. Miyake and T. Koyoma. Maruzen, Tokyo. 13–37.

ICHIMURA, S., S. NAGASAWA and T. TANAKA. 1968. On the oxygen and chlorophyll maxima found in the metalimnion of a mesotrophic lake. *Bot. Mag., Tokyo*, **81**, 1–10.

IKEDA, T. 1970. Relationship between respiration rate and body size in marine plankton animals as a function of the temperature of habitat. *Bull. Fac. Fish. Hokkaido Univ.*, **21**, 91–112.

IKUSHIMA, I. 1967. Ecological studies on the productivity of aquatic plant communities. III. Effect of depth on daily photosynthesis in submerged macrophytes. *Bot. Mag., Tokyo*, **80**, 57–67.

ITOH, K. 1970. A consideration on feeding habits of planktonic copepods in relation to the structure of their oral parts. *Bull. Plankton Soc. Japan*, **17**, 1–10.

IVLEV, V. S. 1944. The time of hunting and the path followed by the predator in relation to the density of the prey population. *Zool. Zh.*, **23** (4), 139–145 (Translation by L. Birkett, Lowestoft).

IVLEV, V. S. 1945. The biological productivity of waters. *Usp. Sovrem. Biol.*, **19**, 98–120.

IVLEV, V. S. 1961. *Experimental Ecology of the Feeding of Fishes*. Translated by D. Scott. Yale Univ. Press, New Haven, 302.

JANNASCH, H. W. 1958. Studies on planktonic bacteria by means of a direct membrane filter method. *J. gen. Microbiol.*, **18**, 609–620.

JANNASCH, H. W. 1960. Versuche über denitrifikation und die Verfügbarkeit des Sauerstoffes in Wassen und Schlamm. *Arch. F. Hydrobiol.*, **56**, 355–369.

JANNASCH, H. W. 1970. Threshold concentrations of carbon sources limiting bacterial growth in sea water. In *Organic matter in Natural Waters*, Ed. D. W. Hood. University of Alaska, 321–328.

JANNASCH, H. W. 1972. New approaches to assessment of microbial activity in polluted waters. In *Water Pollution Microbiology*, Ed. R. Mitchell. J. Wiley, New York, 291–303.

JEFFREY, S. W. 1961. Paper chromatographic separation of chlorophylls and carotenoids from marine algae. *Biochem. J.*, **80**, 336–342.

JEFFREY, S. W. 1969. Properties of two spectrally different components in chlorophyll *c* preparations. *Biochim. Biophys. Acta*, **177**, 456–467.

JEFFREY, S. W., and M. B. ALLEN. 1964. Pigments, growth and photosynthesis in cultures of two chrysomonads, *Coccolithus huxleyi* and a *Hymenomonas* sp. *J. gen. Microbiol.*, **36**, 277–288.

JEFFRIES, H. P. 1962. Environmental characteristics of Raritan Bay, a polluted estuary. *Limnol. Oceanogr.*, **7**, 21–31.

JEFFRIES, H. P. 1969. Seasonal composition of temperate plankton communities: free amino acids. *Limnol. Oceanogr.*, **14**, 41–52.

JEFFRIES, H. P. 1970. Seasonal composition of temperate plankton communities: fatty acids. *Limnol. Oceanogr.*, **15**, 419–426.

JERLOV, N. G. 1957. Optical studies of ocean waters. *Rep. Sweden Deep Sea Exped.*, **3**, 1–59.

JERLOV, N. G. 1968. *Optical Oceanography.* Elsevier, New York, 194.

JITTS, H. R. 1963. The simulated *in situ* measurement of oceanic primary production. *Aust. J. Mar. Freshw. Res.*, **14**, 139–147.

JOHANNES, R. E. 1965. Influence of marine protozoa on nutrient regeneration. *Limnol. Oceanogr.*, **10**, 434–442.

JOHANNES, R. E., and K. L. WEBB. 1965. Release of dissolved amino acids by marine zooplankton. *Science*, **150**, 76–77.

JOHANNES, R. E. 1968. Nutrient regeneration in lakes and oceans. In *Advances in Microbiology of the Sea.* Eds. M. R. Droop and E. J. Ferguson Wood. Academic Press, New York, 203–213.

JOHANNES, R. E., and K. L. WEBB. 1970. Release of dissolved organic compounds by marine and fresh water invertebrates. In *Organic Matter in Natural Waters.* Ed. D. W. Hood. University of Alaska, 257–273.

JOHNSTON, R. 1963. Seawater, the natural medium of phytoplankton. I. General features. *J. mar. biol. Ass. U.K.*, **43**, 427–456.

JONES, D., and M. S. WILLS. 1956. The attenuation of light in sea and estuarine waters in relation to the concentration of suspended solid matter. *J. mar. biol. Ass. U.K.*, **35**, 431–444.

JONES, G. E. 1963. Suppression of bacterial growth by sea water. In *Symposium on Marine Microbiology.* Ed. C. H. Oppenheimer. C. C. Thomas, Springfield, Illinois, 572–579.

JONES, G. E. 1964. Effect of chelating agents on the growth of *Escherichia coli* in sea water. *J. Bacteriol.*, **87**, 483–499.

JONES, G. E. 1967. Precipitates from autoclaved sea water. *Limnol. Oceanogr.*, **13**, 165–167.

JONES, G. E. 1971. The fate of freshwater bacteria in the sea. *Developments in Indust. Microbiol.*, **12**, 141–151.

JØRGENSEN, C. B. 1966. *Biology of Suspension Feeding*, Publ. Pergamon Press, London, 357.

JØRGENSEN, E. G. 1964. Adaptation to different light intensities in the diatom *Cyclostella memeghiniana* Küts. *Physiol. Plant.*, **17**, 136–145.

JØRGENSEN, E. G. 1966. Photosynthetic activity during the life cycle of synchronous *Skeletonema* cells. *Physiol. Plant.*, **19**, 789–799.

KAJIHARA, M. 1971. Settling velocity and porosity of large suspended particles. *J. Oceanogr. Soc. Japan*, **27**, 158–162.

KANE, J. E. 1967. Organic aggregates in the surface waters of the Ligurian Sea. *Limnol. Oceanogr.*, **12**, 287–294.

KAYAMA, M. 1964. Fatty acid metabolism of fishes. *Bull. Japanese Soc. Sci. Fish.*, **30**, 647–659. (In Japanese).

KERR, S. R. 1971. A simulation model of lake trout growth. *J. Fish. Res. Bd. Canada*, **28**, 815–819.

KERR, S. R., and N. V. MARTIN. 1970. Trophic-dynamics of lake trout production systems. In *Marine Food Chains.* Ed. J. H. Steele. Oliver and Boyd, Edinburgh, 365–376.

KETCHUM, B. H. 1939a. The absorption of phosphate and nitrate by illuminated cultures of *Nitzschia closterium. Am. J. Bot.*, **26**, 399–407.

KETCHUM, B. H. 1939b. The development and restoration of deficiencies in the phosphorus and nitrogen composition of unicellular plants. *J. Cell. Comp. Physiol.*, **13**, 373–381.

KETCHUM, B. H. 1967. Phytoplankton nutrients in estuaries. In *Estuaries.* Ed. G. H. Lauff, AAAS, Washington, 329–335.

KHMELEVA, N. N. 1967. Role of radiolarians in the estimation of the primary production in the Red Sea and the Gulf of Aden. *Dokl. Akad. Nauk. SSSR*, **172**, 1430–1433. (In Russian). *Dokl. (Proc.) Acad. Sci. USSR*, **172**, 70–72 (English Transl.).

KIBBY, H. V. 1971. Effect of temperature on the feeding behaviour of *Daphnia rosea. Limnol. Oceanogr.*, **16**, 580–581.

KIEFER, D., and J. D. H. STRICKLAND. 1970. A comparative study of photosynthesis in seawater samples incubated under two types of light attenuator. *Limnol. Oceanogr.*, **15**, 408–412.

KIMATA, M., H. KADOTA, Y. HATA and T. TAJIMA. 1955. Studies on the marine sulfite-reducing bacteria. I. Distribution of marine sulfate-reducing bacteria in the coastal waters receiving a considerable amount of pulp-mill drainage (Japanese, English summary). *Bull. Japan Soc. Sci. Fisheries*, **21**, 102–108.

KOBLENTZ-MISHKE, O. J., V. V. VOLKOVISNKY and J. G. KABANOVA. 1970. Plankton primary production of the world ocean. In *Scientific Exploration of the South Pacific.* Standard Book No. 309-01755-6 Nat. Acad. Sci. Wash., 183–193.

KOK, B., and G. HOCH. 1961. Spectral changes in photosynthesis. In *A Symposium on Light and Life.* Eds. W. D. McElroy and B. Glass, The John Hopkins Press, Baltimore, 397–423.

KORINEK, J. 1927. Ein Beitrag zur Mikrobiologie des Meeres. *Zentralbl. Bakteriol.*, **71**, 73–79.

KRAUSE, H. R. 1961. Einige Bemerkungen über den postmortalen Abbau von Süsswasser—Zooplankton unter laboratoriums—und Freiland bedingungen. *Arch. Hydrobiol.*, **57**, 539–543.

KRAUSE, H. R. 1962. Investigation of the decomposition of organic matter in natural waters. *FAO Fish. Biol. Rep. No. 34*, pp. 19.

KRAUSE, H. R., L. MOCHEL and M. STEGMANN. 1961. Organische Sauren als geloste Intermediarprodukte des postmortalen Abbaues von Süsswasser. *Zooplankton Naturwissenschaften*, **48**, 434–435.

KRISS, A. E. 1963. *Marine Microbiology (Deep Sea)*. Translated by J. M. Shewan and Z. Kabata. Oliver and Boyd, London, pp. 536.

KRISS, A. E., M. N. LEBEDEVA and I. N. MITZKEVICH. 1960. Microorganisms as indicators of hydrological phenomena in seas and oceans. II. Investigation of the deep circulation of the Indian Ocean using microbiological methods. *Deep-Sea Res.*, **6**, 173–183.

KUENEN, P. H. 1950. *Marine Geology*. John Wiley and Sons, New York, pp. 568.

KUENZLER, E. J. 1965. Glucose-6-phosphate utilization by marine algae. *J. Phycol.*, **1**, 156–164.

KUENZLER, E. J. 1970. Dissolved organic phosphorus excretion by marine phytoplankton. *J. Phycol.*, **6**, 7–13.

KUENZLER, E. J., and B. H. KETCHUM. 1962. Rate of phosphorus uptake by *Phaeodactylum tricornutum*. *Biol. Bull.*, **123**, 134–145.

KUZNETSOV, S. I. 1959. *Die Rolle der Mikroorganismen im Stoffkreislauf der Seen*. VEB Deutscher Verlag der Wissenschaffen, Berlin, pp. 301.

KUZNETSOV, S. I. 1968. Recent studies on the role of microorganisms in the cycling of substances in lakes. *Limnol. Oceanogr.*, **13**, 211–224.

LAFOND, E. C., and K. G. LAFOND. 1971. Oceanography and its relation to marine organic production. In *Fertility of the Sea*. Ed. J. D. Costlow, Publ. Gordon and Breach (New York), vol. 1, 241–265.

LANCE, J. 1962. Effects of water of reduced salinity on the vertical migration of zooplankton. *J. mar. biol. Ass. U.K.*, **42**, 131–154.

LARSEN, H. 1962. Halophilism. In *The Bacteria. A Treatise on Structure and Function. IV. The Physiology of Growth*. Eds. I. C. Gunsalus and R. Y. Stanier. Academic Press, New York, 297–342.

LASKER, R. 1960. Utilization of organic carbon by a marine crustacean. Analysis with Carbon-14. *Science*, **131**, 1098–1100.

LASKER, R. 1966. Feeding, growth, respiration and carbon utilization of a euphausiid crustacean. *J. Fish. Res. Bd. Canada*, **23**, 1291–1317.

LEBRASSEUR, R. J. 1969. Growth of juvenile chum salmon (*Oncorhynchus keta*) under different feeding regimes. *J. Fish. Res. Bd. Canada*, **26**, 1631–1645.

LEBRASSEUR, R. J., and J. FULTON. 1967. A guide to zooplankton of the north western Pacific Ocean. *Fish. Res. Bd. Canada, Circular No. 84*, pp. 34.

LEBRASSEUR, R. J., and O. D. KENNEDY. 1972. The fertilization of Great Central Lake. II. Zooplankton standing stock. *Fish. Bull.*, **70**, 25–36.

LEE, R. F., J. C. NEVENZEL and G. A. PAFFENHÖFER. 1970. Wax esters in marine copepods. *Science*, **167**, 1510–1511.

LEE, R. F., J. C. NEVENZEL and G. A. PAFFENHÖFER. 1971. Importance of wax esters and other lipids in the marine food chain: phytoplankton and copepods. *Mar. Biol.*, **9**, 99–108.

LEONG, R. J. H., and C. P. O'CONNELL. 1969. A laboratory study of particulate and filter feeding of northern anchovy (*Engraulis mordas*). *J. Fish. Res. Bd. Canada*, **26**, 557–582.

LEVANIDOV, V. I. 1949. The significance of allochthonous material as food resource in a water-basin and its consumption by *Asellus aquaticus*. *Trudy uses. gidrobiol. Obslch.*, **1**. (In Russian).

LEWIN, J. C. 1957. Silicon metabolism in diatoms. IV. Growth and frustule formation in *Navicula pelliculosa*. *Can. J. Microbiol.*, **3**, 427–433.

LEWIN, J. C., R. A. LEWIN and D. E. PHILPOTT. 1958. Observations on *Phaeodactylum tricornutum*. *J. gen. Microbiol.*, **18**, 418–426.

LEWIN, J. C., and R. A. LEWIN. 1960. Auxotrophy and heterotrophy in marine littoral diatoms. *Can. J. Microbiol.*, **6**, 127–134.

LEWIS, A. G. 1967. An enrichment solution for culturing the early developmental stages of the planktonic marine copepod *Euchaeta japonica*. Marukawa. *Limnol. Oceanogr.*, **12**, 147–148.

LEWIS, A. G., A. RAMNARINE, and M. S. EVANS. 1971. Natural chelators—an indication of activity with the calanoid copepod *Euchaeta japonica*. *Mar. Biol.*, **11**, 1–4.

LEWIS, R. W. 1969. The fatty acid composition of Arctic marine phytoplankton and zooplankton with special reference to minor acids. *Limnol. Oceanogr.*, **14**, 35–40.

LIEBIG, J. 1840. *Chemistry in its application to agriculture and physiology*. Taylor and Walton, London, (4th ed., 1847), pp. 352.

LILLELUND, K. 1965. Effect of abiotic factors in young stages of marine fish. *ICNAF Spec. Publ.*, **6**, 674–686.

LISIVNENKO, L. N. 1961. Plankton and the food of larval Baltic herring, in the Gulf of Riga. Trudy N.—I. *Instituta Rybnogo Khoziaistva Soveta Narodnogo Khoziaistva Latviiskoi SSR*, **3**, 105–138 (Fish. Res. Bd. Canada, Trans. No. 444. pp. 36 1963).

LLOYD, M., and R. J. GHELARDI. 1964. A table for calculating the 'equitability' component of species diversity. *J. Anim. Ecol.*, **33**, 217–225.

LLOYD, M., J. H. ZAR and J. R. KARR. 1968. On the calculation of information—theoretical measures of diversity. *Amer. Midland Nat.*, **79**, 257–272.

LOEBLICH, A. R., III. 1966. Aspects of the physiology and biochemistry of the Pyrrhophyta. *Phykos, Prof. Iyengar Memorial Volume*, **5**, 216–255.

LONGHURST, A. R., A. D. REITH, R. E. BOWER and D. L. R. SEIBERT. 1966. A new system for the collection of multiple serial plankton samples. *Deep-Sea Res.*, **13**, 213–222.

LORENZEN, C. J. 1963. Diurnal variation in photosynthetic activity of natural phytoplankton populations. *Limnol. Oceanogr.*, **8**, 56–62.

LORENZEN, C. J. 1965. A note on the chlorophyll and phaeophytin content of the chlorophyll maximum. *Limnol. Oceanogr.*, **10**, 482–483.

LYMAN, J., and R. H. FLEMING. 1940. Composition of sea water. *J. Mar. Res.*, **3**, 134–146.

MCALLISTER, C. D. 1970. Zooplankton rations, phytoplankton mortality and the estimation of marine production. In *Marine Food Chains*. Ed. J. H. Steele. Oliver and Boyd, Edinburgh, 419–457.

MCALLISTER, C. D., N. SHAH and J. D. H. STRICKLAND. 1964. Marine phytoplankton photosynthesis as a function of light intensity: a comparison of methods. *J. Fish. Res. Bd. Canada*, **21**, 159–181.

MCALLISTER, C. D., R. J. LEBRASSEUR and T. R. PARSONS. 1972. Stability of enriched aquatic ecosystems. *Science*, **175**, 562–564.

MCCARTHY, J. J. 1970. A urease method for urea in seawater. *Limnol. Oceanogr.*, **15**, 309–313.

MCGOWAN, J. A. 1971. Oceanic biogeography of the Pacific. In *The Micropaleontology of Oceans*. Eds. B. M. Funnell and W. R. Riedel. Cambridge University Press, 3–74.

MCINTYRE, A., and A. W. H. BÉ. 1967. Modern Coccolithophoridae of the Atlantic Ocean. I. Placoliths and cyrtoliths. *Deep-Sea Res.*, **14**, 561–597.

MCLAREN, I. A. 1963. Effects of temperature on growth of zooplankton and the adaptive value of vertical migration. *J. Fish. Res. Bd. Canada*, **20**, 685–727.

MCLAREN, I. A. 1965. Some relationships between temperature and egg size, body size, development rate and fecundity of the copepod *Pseudocalanus*. *Limnol. Oceanogr.*, **10**, 528–538.

MCLAREN, I. A. 1966. Predicting development rate of copepod eggs. *Biol. Bull.*, **131**, 457–469.

MCLAUGHLIN, J. J. A., and P. A. ZAHL. 1966. Endozoic algae. In *Symbiosis*, Ed. S. M. Henry, Academic Press, New York, 257–297.

MCNAUGHT, D. C., and A. D. HASLER. 1964. Rate of movement of populations of *Daphnia* in relation to changes in light intensity. *J. Fish. Res. Bd. Canada*, **21**, 291–318.

MACARTHUR, R. H. 1955. Fluctuations of animal populations and a measure of community stability. *Ecology*, **36**, 533–536.

MACISAAC, J. J., and R. C. DUGDALE. 1969. The kinetics of nitrate and ammonia uptake by natural populations of marine phytoplankton. *Deep-Sea Res.*, **16**, 45–57.

MACKINNON, D. L., and R. S. J. HAWES. 1961. *An Introduction to the Study of Protozoa*. Clarendon Press, Oxford, pp. 506.

MACLEOD, R. A. 1965. The question of the existence of specific marine bacteria. *Bact. Rev.*, **29**, 9–23.

MALONE, T. C. 1971. Diurnal rhythms in netplankton and nannoplankton assimilation ratios. *Mar. Biol.*, **10**, 285–289.

MALY, E. J. 1969. A laboratory study of the interaction between the predatory rotifer *Aspalanchina* and *Paramecium*. *Ecology*, **50**, 59–73.

MANN, K. H. 1965. Energy transformations by a population of fish in the River Thames. *J. Anim. Ecol.*, **34**, 253–275.

MANN, K. H. 1969. The dynamics of aquatic ecosystems. *Adv. Ecol. Res.*, **6**, 1–81.

MARGALEF, D. R. 1951. Diversidad de especies en les communidades naturales. *Publ. Inst. Biol. apl., Barcelona*, **9**, 5–27.

MARGALEF, D. R. 1957. La teoria de la informacion en ecologia. *Mem. Real. Acad. Ciencias y Artes de Barcelona*, **23**, 373–449. (Translation by Wendell Hall, General Systems, Yearbook, 3, 36–71, 1957.)

MARGALEF, D. R. 1958. *Temporal succession and spatial heterogeneity in phytoplankton. Perceptives in Marine Biology*. Ed. A. A. Buzzati-Traverso, University California Press, Berkeley, 323–349.

MARGALEF, D. R. 1961. Correlations entre certains caractères synthétiques des populations de phytoplankton. *Hydrobiologia*, **18**, 155–164.

MARGALEF, D. R. 1965. Ecological correlations and the relationship between primary productivity and community structure. In *Primary Productivity in Aquatic Environments*. Ed. C. H. Goldman. University of California Press, Berkeley. *Mem. Ist. Ital. Idrobiol.*, **18** Suppl., 355–364.

MARSHALL, P. T. 1958. Primary production in the Arctic. *J. Cons. Int. Explor. Mer*, **23**, 173–177.

MARSHALL, S. M., and A. P. ORR. 1955. *The biology of a marine copepod, Calanus finmarchicus* (Gunnerus). Oliver and Boyd, Edinburgh, pp. 188.

MARTIN, J. H. 1968. Phytoplankton—zooplankton relationships in Narragansett Bay. III. Seasonal changes in zooplankton excretion rates in relation to phytoplankton abundance. *Limnol. Oceanogr.*, **13**, 63–71.

MARUMO, R., N. TAGA and T. NAKAI. 1971. Neustonic bacteria and phytoplankton in surface microlayers of the equatorial waters. *Bull. Plankton Soc. Japan*, **18**, 36–41.

MELAND, S. M. 1962. Marine alginate-decomposing bacteria from north Norway. *Nytt Mag. Bot.*, **10**, 53–80.

MENZEL, D. W. 1964. The distribution of dissolved organic carbon in the Western Indian Ocean. *Deep-Sea Res.*, **11**, 757–765.

MENZEL, D. W. 1966. Bubbling of sea water and the production of organic particles: a re-evaluation. *Deep-Sea Res.*, **13**, 963–966.

MENZEL, D. W. 1967. Particulate organic carbon in the deep sea. *Deep-Sea Res.*, **14**, 229–238.

MENZEL, D. W., and J. H. RYTHER. 1960. The annual cycle of primary production in the Sargasso Sea off Bermuda. *Deep-Sea Res.*, **6**, 351–367.

MENZEL, D. W., and J. H. RYTHER. 1961. Nutrients limiting the production of phytoplankton in the Sargasso Sea, with special reference to iron. *Deep-Sea Res.*, **7**, 276–281.

MENZEL, D. W., E. M. HULBURT and J. H. RYTHER. 1963. The effects of enriching Sargasso Sea water on the production and species composition of the phytoplankton. *Deep-Sea Res.*, **10**, 209–219.

MENZEL, D. W., and J. H. RYTHER. 1964. The composition of particulate organic matter in the western North Atlantic. *Limnol. Oceanogr.*, **9**, 179–186.

MENZEL, D. W., and J. J. GOERING. 1966. The distribution of organic detritus in the ocean. *Limnol. Oceanogr.*, **11**, 333–337.

MENZEL, D. W., J. ANDERSON and A. RADTKE. 1970. Marine phytoplankton vary in their response to chlorinated hydrocarbons. *Science*, **167**, 1724–1726.

MINDERMAN, G. 1968. Addition, decomposition and accumulation of organic matter in forests. *J. Ecology*, **56**, 335–362.

MITCHELL, R. 1968. Factors affecting the decline of non-marine microorganisms in seawater. *Water. Res.*, **2**, 535–543.

MORRIS, I., and P. J. SYRETT. 1963. The development of nitrate reductase in *Chlorella* and its repression by ammonium. *Arch. Mikrobiol.*, **47**, 32–41.

MOSHKINA, L. V. 1961. Photosynthesis by dinoflagellatae of the Black Sea. *Fiziologiya Rastenii*, **8**, 172–177 (In Russian). *Plant. Physiol.*, **8**, 129–132. (English translation.)

MULLIN, M. M. 1963. Some factors affecting the feeding of marine copepods of the genus *Calanus*. *Limnol. Oceanogr.*, **8**, 239–250.

MULLIN, M. M., and E. R. BROOKS. 1967. Laboratory culture, growth rate, and feeding behavior of a planktonic marine copepod. *Limnol. Oceanogr.*, **12**, 657–666.

MULLIN, M. M., and E. R. BROOKS. 1970. Growth and metabolism of two planktonic marine copepods as influenced by temperature and type of food. In *Marine Food Chains*. Ed. J. H. Steele. Oliver and Boyd, Edinburgh, 74–95.

MUNK, W. H., and G. A. RILEY. 1952. Absorbtion of nutrients by aquatic plants. *J. mar. Res.*, **11**, 215–240.

MYERS, J. 1953. In *The Metabolism of Algae*, by G. E. Fogg. Methuen, London, pp. 149.

NAKAJIMA, K., and S. NISHIZAWA. 1968. Seasonal cycles of chlorophyll and seston in the surface water of the Tsugaru Strait area. *Records Oceanogr. Works, Japan*, **9**, 219–246.

NAKANISHI, M., and M. MONJI. 1965. Effect of variation in salinity on photosynthesis of phytoplankton growing in estuaries. *J. Fac. Sci., Univ. Tokyo, Sec. III Botany*, **9**, 19–42.

NASSOGNE, A. 1970. Influence of food organisms on the development and culture of pelagic copepods. *Helgoländer wiss. Meeresunters.*, **20**, 333–345.

NEES, J. C. 1949. *A contribution to aquatic population dynamics*. Ph.D. Thesis, University of Wisconsin, Madison.

NEMOTO, T. 1967. Feeding pattern of euphausiids and differentiations in their body characteristics. *Inf. Bull. Plankton. Japan* (Comm. No. Dr. Y. Matsue), 157–171.

NEMOTO, T., and K. ISHIKAWA. 1969. Organic particulate and aggregate matters stained by histological reagents in the East China Sea. *J. Oceanogr. Soc. Japan*, **25**, 281–290.

NEWELL, B. S. 1967. The determination of ammonia in seawater. *J. mar. biol. Ass. U.K.*, **47**, 271–280.

NEWELL, G. E., and R. C. NEWELL. 1963. *Marine plankton*. Hutchinson Educational Ltd., London. (Revised edition). 1966, pp. 221.

NEWELL, R. 1965. The role of detritus in the nutrition of two marine deposit feeders, the prosobranch *Hydrobia ulvae* and the bivalve *Macoma balthica*. *Proc. Zool. Soc. London*, **144**, 25–45.

NIHEI, T., T. SASA, S. MIYACHI, K. SUZUKI and H. TAMIYA. 1954. Change of photosynthetic activity of *Chlorella* cells during the course of their normal life cycle. *Archiv. Mikrobiol.*, **21**, 156–166.

NISHIZAWA, S. 1969. Suspended material in the sea. II. Re-evaluation of the hypotheses. *Bull. Plankton Soc. Japan*, **16**, 1–42.

NISHIZAWA, S. 1971. Concentration of organic and inorganic material in the surface skin at the equator, 155°W. *Bull. Plankton Soc. Japan*, **18**, 42–44.

NORTH, B. B., and G. C. STEPHENS. 1971. Uptake and assimilation of amino acids by *Platymonas*. II. Increased uptake in nitrogen-deficient cells. *Biol. Bull.*, **140**, 242–254.

NORTH, B. B., and G. C. STEPHENS. 1972. Amino acid transport in *Nitzschia ovalis* Arnott. *J. Phycol.*, **8**, 64–68.

ODUM, E. P., and A. A. DE LA CRUZ. 1963. Detritus as a major component of ecosystems. *A.I.B.S. Bulletin*, **13**, 39–40.

'OHEOCHA, C., and M. RAFTERY. 1959. Phycoerythrins and phycocyanins of cryptomonads. *Nature*, **184**, 1049–1051.

OLSON, J. S. 1963. Energy storage and the balance of producers and decomposers in ecological systems. *Ecology*, **44**, 322–331.

OMORI, M. 1965. The distribution of zooplankton in the Bering Sea and northern North Pacific, as observed by high-speed sampling of the surface waters, with special reference to the copepods. *J. Oceanogr. Soc. Japan*, **21**, 18–27.

OMORI, M. 1969. Weight and chemical composition of some important oceanic zooplankton in the North Pacific Ocean. *Mar. Biol.*, **3**, 4–10.

OSTERBERG, C., A. G. CAREY and H. CURL. 1963. Acceleration of sinking rates of radionuclides in the ocean. *Nature*, **200**, 1276–1277.

OTSUKI, A., and T. HANYA. 1968. On the production of dissolved nitrogen-rich organic matter. *Limnol. Oceanogr.*, **13**, 183–185.

OWEN, R. W., and B. ZEIZSCHEL. 1970. Phytoplankton production: seasonal changes in the oceanic eastern tropical Pacific. *Mar. Biol.*, **7**, 32–36.

PAASCHE, E. 1962. Coccolith formation. *Nature*, **193**, 1094–1095.

PAASCHE, E. 1966. Action spectrum of coccolith formation. *Physiol. Plant*, **19**, 770–779.

PAFFENHÖFER, G. A. 1970. Cultivation of *Calanus helgolandicus* under controlled conditions. *Helgoländer wiss. Meeresunters*, **20**, 346–359.

PAFFENHÖFER, G. A., and J. D. H. STRICKLAND. 1970. A note on the feeding of *Calanus helgoalndicus* on detritus. *Mar. Biol.*, **5**, 97–99.

PALOHEIMO, J. E., and L. M. DICKIE. 1965. Food and growth of fishes. I. A growth curve derived from experimental data. *J. Fish. Res. Bd. Canada*, **22**, 521–542.

PALOHEIMO, J. E., and L. M. DICKIE. 1966a. Food and growth of fishes. II. Effects of food and temperature on the relation between metabolism and body weight. *J. Fish. Res. Bd. Canada*, **23**, 869–908.

PALOHEIMO, J. E., and L. M. DICKIE. 1966b. Food and growth of fishes. III. Relation among food, body size and growth efficiency. *J. Fish. Res. Bd. Canada*, **23**, 1209–1248.

PARK, K. 1967. Nutrient regeneration of preformed nutrients off Oregon. *Limnol. Oceanogr.*, **12**, 353–357.

PARKE, M., and P. S. DIXON. 1968. Check-list of British Marine algae—second revision. *J. mar. biol. Ass. U.K.*, **48**, 783–832.

PARKER, P. L., C. VAN BAALEN and L. MAURER. 1967. Fatty acids in eleven species of blue-green algae: geochemical significance. *Science*, **155**, 707–708.

PARKER, R. R. 1971. Size selective predation among juvenile salmonid fishes in a British Columbia inlet. *J. Fish. Res. Bd. Canada*, **28**, 1503–1510.

PARSONS, T. R. 1961. On the pigment composition of eleven species of marine phytoplankters. *J. Fish. Res. Bd. Canada*, **18**, 1017–1025.

PARSONS, T. R. 1969. The use of particle size spectra in determining the structure of a plankton community. *J. Oceanogr. Soc. Japan*, **25**, 172–181.

PARSONS, T. R., K. STEPHENS and J. D. H. STRICKLAND. 1961. On the chemical composition of eleven species of marine phytoplankters. *J. Fish. Res. Bd. Canada*, **18**, 1001–1016.

PARSONS, T. R., and J. D. H. STRICKLAND. 1962a. Oceanic detritus. *Science*, **136**, 313–314.

PARSONS, T. R., and J. D. H. STRICKLAND. 1962b. On the production of particulate organic carbon by heterotrophic processes in sea water. *Deep-Sea Res.*, **8**, 211–222.

PARSONS, T. R., R. J. LeBRASSEUR and J. D. FULTON. 1967. Some observations on the dependence of zooplankton grazing on the cell size and concentration of phytoplankton blooms. *J. Oceanogr. Soc. Japan*, **23**, 10–17.

PARSONS, T. R., R. J. LeBRASSEUR, J. D. FULTON and O. D. KENNEDY. 1969. Production studies in the Strait of Georgia. Part II. Secondary production under the Fraser River plume, February to May, 1967. *J. Exp. Mar. Biol. Ecol.*, **3**, 39–50.

PARSONS, T. R., and R. J. LeBRASSEUR. 1970. The availability of food to different trophic levels in the marine food chain. In *Marine Food Chains*. Ed. J. H. Steele. Oliver and Boyd, Edinburgh, 325–343.

PARSONS, T. R., and H. SEKI. 1970. Importance and general implications of organic matter in aquatic environments. In *Organic Matter in Natural Waters*. Ed. D. W. Hood. University of Alaska, 1–27.

PARSONS, T. R., K. STEPHENS and M. TAKAHASHI. 1972. The fertilization of Great Central Lake. I. Effect of primary production. *Fish. Bull.*, **70**, 13–23.

PATTEN, B. C. 1959. An introduction to the cybernetics of the ecosystem: the trophic-dynamic aspect. *Ecology*, **40**, 221–231.

PATTEN, B. C. 1961. Preliminary method for estimating stability in plankton. *Science*, **134**, 1010–1011.

PATTEN, B. C. 1962a. Species diversity in net phytoplankton of Raritan Bay. *J. Mar. Res.*, **20**, 57–75.

PATTEN, B. C. 1962b. Improved method for estimating stability in plankton. *Limnol. Oceanogr.*, **7**, 266–268.

PATTEN, B. C. 1968. Mathematical models of plankton production. *Int. Rev. ges. Hydrobiol.*, **53**, 357–408.

PATTON, S., P. T. CHANDLER, E. B. KALAN, A. R. LOEBLICH III, G. FULLER and A. A. BENSON. 1967. Food value of red tide (*Gonyaulax polyedra*). *Science*, **158**, 789–790.

PETIPA, T. S. 1966. Relationship between growth, energy metabolism and ration in *Acartia clausi* Giesbr. Physiology of marine animals. Akademiya Nauk SSSR. *Oceanogr. Comm.* 82–91 (translated by M. A. Paranjape, U. Wash.).

PETIPA, T. S., E. V. PAVLOVA and G. N. MIDONOV. 1970. The food web structure, utilization and transport of energy by trophic levels in the planktonic communities. *In Marine Food Chains*. Ed. J. H. Steele. Oliver and Boyd, Edinburgh, 142–167.

PICKARD, G. L. 1964. *Descriptive Physical Oceanography*. Pergamon Press, Oxford, pp. 199.

PIELOU, E. C. 1966. The measurement of diversity in different types of biological collections. *J. Theoret. Biol.*, **13**, 131–144.

PLATT, T. 1969. The concept of energy efficiency in primary production. *Limnol. Oceanogr.*, **14**, 653–659.

PLATT, T. 1971. The annual production by phytoplankton in St. Margaret's Bay, Nova Scotia. *J. Cons. Int. Explor. Mer*, **33**, 324–334.

PLATT, T., V. M. BRAWN and B. IRWIN. 1969. Caloric and carbon equivalents of zooplankton biomass. *J. Fish. Res. Bd. Canada*, **26**, 2345–2349.

PLATT, T., L. M. DICKIE and R. W. TRITES 1970. Spatial heterogeneity of phytoplankton in a near-shore environment. *J. Fish. Res. Bd. Canada*, **27**, 1453–1473.

POMEROY, L. R., H. M. MATHEWS and H. S. MIN. 1963. Excretion of phosphate and soluble organic phosphorus compounds by zooplankton. *Limnol. Oceanogr.*, **8**, 50–55.

PORTER, J. R. 1946. *Bacterial Chemistry and Physiology*. Chapman and Hall, London, pp. 1073.

PRAKASH, A. 1971. Terrigenous organic matter and coastal phytoplankton fertility. In *Fertility of the Sea*. Ed. J. D. Costlow. Gordon and Breach, New York, vol. 2, 351–368.

PRAKASH, A., and M. A. RASHID. 1968. Influence of humic substances on the growth of marine phytoplankton: dinoflagellates. *Limnol. Oceanogr.*, **13**, 598–606.

PRATT, D. W. 1966. Competition between *Skeletonema costatun* and *Olithodiscus luteus* in Narragansett Bay and in culture. *Limnol. Oceanogr.*, **11**, 447–455.

PROVASOLI, L. 1958. Nutrition and ecology of protozoa and algae. *Ann. Rev. Microbiol.*, **12**, 279–308.

PROVASOLI, L., J. J. A. McLAUGHLIN and M. R. DROOP. 1957. The development of artificial media for marine algae. *Archiv. Mikrobiol.*, **25**, 392–428.

PROVASOLI, L., K. SHIRAISHI and J. R. LANCE. 1959. Nutritional idiosyncrasies of *Artemia* and *Tigriopus* in monoxenic culture. *Ann. New York Acad. Sci.*, **77**, 250–261.

PROVASOLI, L., and J. J. A. McLAUGHLIN. 1963. Limited heterotrophy of some photosynthetic dinoflagellates. In *Symposium on Marine Microbiology*. Ed. C. H. Oppenheimer. C. C. Thomas, Springfield, Illinois, 105–113.

PSHENIN, L. N. 1963. Distribution and ecology of *Azotobacter* in the Black Sea. In *Symposium on Marine Microbiology*. Ed. C. H. Oppenheimer. C. C. Thomas, Springfield, Illinois, 383–391.

RABINOWITCH, E. I. 1951. *Photosynthesis and related processes*. Vol. II, Part I. Interscience, New York, 1211–2088.

RAE, K. M. 1957. A relationship between wind, plankton distribution and haddock brood strength. *Bull. Mar. Ecol.*, **4**, 347–369.

RAYMONT, J. E. G., and R. J. CONOVER. 1961. Further investigations on the carbohydrate content of marine zooplankton. *Limnol. Oceanogr.*, **6**, 154–164.

RAYMONT, J. E. G., R. T. SRINIVASAGAM and J. K. B. RAYMONT. 1969a. Biochemical studies on marine zooplankton. VII. Observations on certain deep sea zooplankton. *Int. Rev. ges. Hydrobiol.*, **54**, 357–365.

RAYMONT, J. E. G., R. T. SRINIVASAGAM and J. K. B. RAYMONT. 1969b. Biochemical studies on marine zooplankton. VI. Investigations on *Meganyctiphanes norvegica* (M. Sars). *Deep-Sea Res.*, **16**, 141–156.

REDFIELD, A. C. 1934. On the proportions of organic derivatives in sea water and their relation to the composition of plankton. *James Johnstone Memorial Volume (Liverpool)*, pp. 176.

REDFIELD, A. C. 1942. The processes determining the concentration of oxygen, phosphate and other organic derivatives within the depths of the Atlantic Ocean. *Pap. Phys. Oceanogr. Met.*, **9**, 1–22.

REDFIELD, A. C. 1955. The hydrography of the Gulf of Venezuela. Papers Marine Biol. Oceanogry., *Deep-Sea Res. Suppl.* **3**, 115–133.

REEVE, M. R. 1963. The filter-feeding of *Artemia*. I. In pure cultures of plant cells. *J. Exptl. Biol.*, **40**, 195–205.

REEVE, M. R. 1964. Feeding of zooplankton, with special reference to some experiments with *Sagitta*. *Nature*, **201**, 211–213.

REMSEN, C. C. 1971. The distribution of urea in coastal and oceanic waters. *Limnol. Oceanogr.*, **16**, 732–740.

Report of the Committee on Terms and Equivalents. 1958. *Rapp. Proces-Verbaux. Reunions*, **144**, 15–16.

RICHMAN, S. 1958. The transformation of energy by *Daphnia pulex*. *Ecol. Monogr.*, **28**, 273–291.

RICHMAN, S., and J. N. ROGERS. 1969. The feeding of *Calanus helgolandicus* on synchronously growing populations of the marine diatom *Ditylium brightwelli*. *Limnol. Oceanogr.*, **14**, 701–709.

RICKER, W. E. 1937. Statistical treatment of sampling processes useful in the enumeration of plankton organisms. *Arch. Hydrobiol. (Plankt.)*, **31**, 68–84.

RICKER, W. E. 1958. Handbook of computations for biological statistics of fish populations. *Fish. Res. Bd. Bull.*, **119**, pp. 300.

RICKETTS, T. R. 1966a. On the chemical composition of some unicellular algae. *Phytochem.*, **5**, 67–76.

RICKETTS, T. R. 1966b. The carotenoids of the phytoflagellate, *Micromonas pusilla*. *Phytochem.*, **5**, 571–580.

RICKETTS, T. R. 1967a. The pigment composition of some flagellates possessing scaly flagella. *Phytochem.*, **6**, 669–676.

RICKETTS, T. R. 1967b. Further investigations into the pigment composition of green flagellates possessing scaly flagella. *Phytochem.*, **6**, 1375–1386.

RICKETTS, T. R. 1970. The pigments of the Prasinophyceae and related organisms. *Phytochem.*, **9**, 1835–1842.

RIGLER, F. H. 1971. Feeding rates. Zooplankton. In *A Manual on methods for the Assessment of Secondary Productivity in Fresh Waters*. Eds. W. T. Edmondson and G. G. Winberg. Blackwell Scientific, Edinburgh, 228–256.

RILEY, G. A. 1946. Factors controlling phytoplankton populations on Georges Bank. *J. Mar. Res.*, **6**, 54–73.

RILEY, G. A. 1947. A theoretical analysis of the zooplankton population of Georges Bank. *J. Mar. Res.*, **6**, 104–113.

RILEY, G. A. 1956. Oceanography of Long Island Sound, 1952–54. II. Physical Oceanography. *Bull. Bingham Oceanogr. Coll.*, **15**, 15–46.

RILEY, G. A. 1963. Organic aggregates in sea water and the dynamics of their formation and utilization. *Limnol. Oceanogr.*, **8**, 372–381.

RILEY, G. A. 1970. Particulate and organic matter in sea water. *Adv. Mar. Biol.*, **8**, 1–118.

RILEY, G. A., H. STOMMEL and D. A. BUMPUS. 1949. Quantitative ecology of the plankton of the western North Atlantic. *Bull. Bingham Oceanogr. Coll.*, **12**, 1–169.

RILEY, J. P., and T. R. S. WILSON. 1967. The pigments of some marine phytoplankton species. *J. mar. biol. Ass. U.K.*, **47**, 351–362.

RILEY, J. P., and D. A. SEGAR. 1969. The pigments of some further marine phytoplankton species. *J. mar. biol. Ass. U.K.*, **49**, 1047–1056.

RILEY, J. P., and I. ROTH. 1971. The distribution of trace elements in some species of phytoplankton grown in culture. *J. mar. biol. Ass. U.K.*, **51**, 63–72.

RILEY, J. P., and R. CHESTER. 1971. *Introduction to marine chemistry*. Academic Press, London, pp. 465.

RISEBROUGH, R. W., D. W. MENZEL, D. J. MARTIN and H. S. OLCOTT. 1967. DDT residues in Pacific sea birds: a persistent insecticide in marine food chains. *Nature*, **216**, 589–591.

ROMANENKO, W. I. 1964a. Potential capacity of the microflora of sludge sediments for heterotrophic assimilation of carbon dioxide and for chemosynthesis. *Mikrobiologiya*, **33**, 134–139 (In Russian).

ROMANENKO, W. I. 1964b. Heterotrophic assimilation of CO_2 by the aquatic microflora. *Mikrobiologiya*, **33**, 679–683 (In Russian).

ROMANENKO, W. I. 1964c. The relationship between the amounts of the O_2 and CO_2 required by heterotrophic bacteria. *Dokl. Akad. Nauk. Sci. SSSR*, **157**, 178–179 (In Russian). *Dokl. (Proc.) Acad. Sci. USSR*, **157**, 562–563 (English transl.).

ROSENTHAL, H., and G. HEMPEL. 1970. Experimental studies in feeding and food requirements of herring larvae (*Clupea harengus* L.). In *Marine Food Chains*. Ed. J. H. Steele. Oliver and Boyd, Edinburgh, 344–364.

ROSENZWEIG, M. L., and R. H. MACARTHUR. 1963. Graphical representation and stability conditions of predator-prey interactions. *Amer. Nat.*, **97**, 209–223.

ROWE, G. T. 1971. Benthic biomass and surface productivity. In *Fertility of the Sea*, Ed. J. D. Costlow. Gordon and Breach, New York, 441–454.

RYTHER, J. H. 1956. Photosynthesis in the ocean as a function of light intensity. *Limnol. Oceanogr.*, **1**, 61–70.

RYTHER, J. H. 1963. IV. Biological Oceanography. 17. Geographic variations in productivity. In *The Sea* Ed. M. N. Hill, Interscience Publishers, New York, vol. 2, 347–380.

RYTHER, J. H. 1969. Photosynthesis and fish production in the sea. The production of organic matter and its conversion to higher forms of life vary throughout the world ocean. *Science*, **166**, 72–76.

RYTHER, J. H., and D. D. KRAMER. 1961. Relative iron requirement of some coastal and off-shore plankton algae. *Ecology*, **42**, 444–446.

RYTHER, J. H., and R. R. L. GUILLARD. 1962. Studies of marine planktonic diatoms. III. Some effects of temperature on respiration of five species. *Can. J. Microbiol.*, **8**, 447–453.

RYTHER, J. H., and D. W. MENZEL. 1965. On the production, composition, and distribution of organic matter in the Western Arabian Sea. *Deep-Sea Res.*, **12**, 199–209.

SAIJO, Y. 1969. Chlorophyll pigments in the deep sea. *Bull. Japanese Soc. Fish. Oceanogr.* Special No. (Prof. Uda's Commemorative Papers), 179–182.

SAIJO, Y., and S. ICHIMURA. 1962. Some considerations on photosynthesis of phytoplankton from the point of view of productivity measurement. *J. Oceanogr. Soc. Japan, 20th Anniv. Vol.*, 687–693.

SAIJO, Y., and K. TAKESUE. 1965. Further studies on the size distribution of photosynthesizing phytoplankton in the Indian Ocean. *J. Oceanogr. Soc. Japan*, **20**, 10–17.

SAKAMOTO, M. 1966. The chlorophyll amount in the euphotic zone in some Japanese lakes and its significance in the photosynthetic production of phytoplankton community. *Bot. Mag., Tokyo*, **79**, 77–88.

SANDERS, H. L. 1968. Marine benthic diversity: a comparative study. *Amer. Nat.*, **102**, 243–282.

SCHAEFER, M. B. 1965. The potential harvest of the sea. *Trans. Amer. Fish. Soc.*, **94**, 123–128.

SEIWELL, H. R., and G. E. SEIWELL. 1938. The sinking of decomposing plankton in sea water and its relationship to oxygen consumption and phosphorus liberation. *Proc. Amer. phil. Soc.*, **78**, 465–481.

SEKI, H. 1965a. Microbial studies on the decomposition of chitin in the marine environment. IX. Rough estimation of chitin decomposition in the ocean. *J. Oceanogr. Soc. Japan*, **21**, 253–260.

SEKI, H. 1965b. Decomposition of chitin in marine sediments. *J. Oceanogr. Soc. Japan*, **1**, 261–268.

SEKI, H. 1968. Relation between production and mineralization of organic matter in Aburatsubo Inlet, Japan. *J. Fish. Res. Bd. Canada*, **25**, 625–637.

SEKI, H. 1972. Formation of Anoxic Zones in Seawater. In *Biological Oceanography of the Northern North Pacific Ocean*. Eds. Y. Takenouchi *et al.*, Idemitsu Shoten, Tokyo, 487–493.

SEKI, H., J. SKELDING and T. R. PARSONS. 1968. Observations on the decomposition of a marine sediment. *Limnol. Oceanogr.*, **13**, 440–447.

SEKI, H., K. V. STEPHENS and T. R. PARSONS. 1969. The contribution of allochthonous bacteria and organic materials from a small river into a semi-enclosed sea. *Arch. Hydrobiol.*, **66**, 37–47.

SHANNON, C. E., and W. WEAVER. 1963. *The Mathematical Theory of Communication*. University of Illinois Press, Urbana. pp. 125.

SHELBOURNE, J. E. 1957. The feeding and condition of plaice larvae in good and bad plankton patches. *J. mar. biol. Ass. U.K.*, **36**, 539–552.

SHELDON, R. W., and T. R. PARSONS. 1967a. *A practical manual on the use of the Coulter Counter in marine science*. Coulter Electronics Sales Co. Canada, Toronto, pp. 66.

SHELDON, R. W., and T. R. PARSONS. 1967b. A continuous size spectrum for particulate matter in the sea. *J. Fish. Res. Bd. Canada*, **24**, 909–915.

SHELDON, R. W., T. P. T. EVELYN and T. R. PARSONS. 1967. On the occurrence and formation of small particles in sea water. *Limnol. Oceanogr.*, **12**, 367–375.

SHELDON, R. W., and W. H. SUTCLIFFE, JR. 1969. Retention of marine particles by screens and filters. *Limnol. Oceanogr.*, **14**, 441–444.

SHIMADA, B. M. 1958. Diurnal fluctuations in photosynthetic rate and chlorophyll *a* content of phytoplankton from eastern Pacific waters. *Limnol. Oceanogr.*, **3**, 336–339.

SIEBURTH, J. McN. 1960. Acrylic acid, an "antibiotic" principle in *Phaeocystis* blooms in Antarctic waters. *Science*, **132**, 676–677.

SIEBURTH, J. McN. 1961. Antibiotic properties of acrylic acid, a factor in the gastrointestinal antibiosis of polar marine animals. *J. Bacteriol.*, **82**, 72–79.

SIEBURTH, J. McN. 1964. Antibacterial substances produced by marine algae. In *Developments in Industrial Microbiology*. Soc. Industr. Microbiol., Washington, D.C., 124–134.

SIEBURTH, J. McN. 1968. Observations on bacteria planktonic in Narragansett Bay, Rhode Island; A résumé. Proceedings of the U.S.–Japan seminar on marine microbiology, *Bull. Misaki Marine Biol. Inst.*, *Kyoto Univ.*, **12**, 49–64.

SIEBURTH, J. McN. 1969. Studies on algal substances in the sea. III. The production of extracellular organic matter by littoral marine algae. *J. Exp. Mar. Biol. Ecol.*, **3**, 290–309.

SIEBURTH, J. McN. 1971. Distribution and activity of oceanic bacteria. *Deep-Sea Res.*, **18**, 1111–1121.

SIEBURTH, J. McN., and A. JENSEN. 1969. Studies on algal substances in the sea. II. The formation of Gelbstoff (humic material) by exudates of Phaeophyta. *J. Exp. Mar. Biol. Ecol.*, **3**, 275–289.

SKOPINTSEV, B. A. 1966. Some aspects of the distribution and composition of organic matter in the waters of the ocean. *Oceanology*, **6**, 441–450. (Fish. Res. Bd. Canada, Transl. No. 930).

SLOAN, P. R., and J. D. H. STRICKLAND. 1966. Heterotrophy of four marine phytoplankters at low substrate concentrations. *J. Phycol.*, **2**, 29–32.

SLOBODKIN, L. B. 1961. *Growth and Regulation of Animal Populations* (Ch. 12). Holt, Rinehart, and Winston, New York, pp. 184.

SMAYDA, T. J. 1958. Biogeographical studies of marine phytoplankton. *Oikos*, **9**, 158–191.

SMAYDA, T. J. 1963. Succession of phytoplankton and the ocean as an holocoenotic environment. In *Symposium on Marine Microbiology*. Ed. C. H. Oppenheimer. C. C. Thomas, Springfield, Illinois, 260–274.

SMAYDA, T. J. 1964. Enrichment experiments using the marine centric diatom *Cyclotella nana* (Clone 13-1) as an assay organism. In *Proceedings of Symposium on experimental marine ecology*, Occasional Publication No. 2, Graduate School of Oceanography, University of Rhode Island, 25–32.

SMAYDA, T. J. 1969. Some measurements of the sinking rate of fecal pellets. *Limnol. Oceanogr.*, **14**, 621–625.

SMAYDA, T. J. 1970a. The suspension and sinking of phytoplankton in the sea. *Oceanogr. Mar. Biol. Ann. Rev.*, **8**, 353–414.

SMAYDA, T. J. 1970b. Growth potential bioassay of water masses using diatom cultures: Phosphorescent Bay (Puerto Rico) and Caribbean Waters. *Helgoländer wiss. Meeresunters.*, **20**, 172–194.

SMAYDA, T. J., and B. J. BOLEYN. 1966. Experimental observations on the flotation of marine diatoms. II *Skeletonema costatum* and *Rhizosolenia setigera*. *Limnol. Oceanogr.*, **11**, 18–34.

SMITH, E. L. 1936. Photosynthesis in relation to light and carbon dioxide. *Proc. Nat. Acad. Science, Wash.*, **22**, 504–511.

SMITH, F. E., and E. R. BAYLOR. 1953. Color responses in the Cladocera and their ecological significance. *Amer. Nat.*, **87**, 49–55.

SOKOLOVA, G. A., and G. I. KARAVAIKO. 1964. *Physiology and geochemical activity of thiobacilli*, Akademiya Nauk SSSR, Institut Mikrobiologii, Moskva [Translated from Russian, Israel Program for Scientific Translation (Jerusalem) 1968], pp. 283.

SOLÓRZANO, L., and J. D. H. STRICKLAND. 1968. Polyphosphate in sea water. *Limnol. Oceanogr.*, **13**, 515–518.

SOROKIN, YU. I. 1961. Heterotrophic carbon dioxide assimilation by micro-organisms. *Zhurnal Obshchei Biologii*, **22**, 265–272 (In Russian).

SOROKIN, YU. I. 1964a. On the trophic role of chemosynthesis in water bodies. *Int. Rev. ges. Hydrobiol.*, **49**, 307–324.

SOROKIN, YU. I. 1964b. A quantitative study of the microflora in the central Pacific Ocean. *J. Cons. Int. Explor. Mer*, **29**, 25–40.

SOROKIN, YU. I. 1964c. On the primary production and bacterial activities in the Black Sea. *J. Cons. Int. Explor. Mer*, **29**, 41–60.

SOROKIN, YU. I. 1966. On the carbon dioxide uptake during the cell synthesis by microorganisms. *Zeitschrift Allg. Mikrobiol.*, **6**, 69–73.

SOROKIN, YU. I. 1969. On the trophic role of chemosynthesis and bacterial biosynthesis in water bodies. In *Primary Productivity in Aquatic Environments.* Ed. C. R. Goldman. Mem. Ist. Ital. Idrobiol., 18 Suppl., University of California Press, Berkeley, 187–205.

SOROKIN, YU. I. 1970. Determination of the activity of heterotrophic microflora in the ocean using C^{14} containing organic matter. *Mikrobiologiya,* **39,** 149–156 (In Russian).

SOROKIN, YU. I. 1971. On the role of bacteria in the productivity of tropical oceanic waters. *Int. Rev. ges. Hydrobiol.,* **56,** 1–48.

SOURNIA, M. A. 1967. Rythme nycthéméral du rapport "intensité photosynthétique chlorophylle" dans le plancton marin. *C. R. Acad. Sc. Paris,* **265,** 1000–1003.

SOURNIA, M. A. 1969. Cycle annuel du phytoplancton et de la production primaire dans les mers tropicales. *Mar. Biol.,* **3,** 287–303.

SPENCER, C. P. 1954. Studies on the culture of a marine diatom. *J. mar. biol. Ass. U.K.,* **33,** 265–290.

SPOEHR, H. A., and H. W. MILNER. 1949. The chemical composition of *Chlorella,* effect of environmental conditions. *Plant Physiol.,* **24,** 120–149.

STARR, T. J. 1956. Relative amounts of vitamin B_{12} in detritus from oceanic and estuarine environments near Sapelo Island, Georgia. *Ecology,* **37,** 658–664.

STAVN, R. H. 1971. The horizontal-vertical distribution hypothesis: Langmuir circulation and *Daphnia* distributions. *Limnol. Oceanogr.,* **16,** 453–466.

STEELE, J. H. 1962. Environmental control of photosynthesis in the sea. *Limnol. Oceanogr.,* **7,** 137–150.

STEELE, J. H. 1964. Some problems in the study of marine resources. *ICNAF Environ. Symp. Rome,* 1964. Contrib. No. C-4, pp. 11.

STEELE, J. H., and C. S. YENTSCH 1960. The vertical distribution of chlorophyll. *J. mar. biol. Ass. U.K.,* **39,** 217–226.

STEEMANN NIELSEN, E. 1952. The use of radioactive carbon (C^{14}) for measuring organic production in the sea. *J. Cons. Int. Explor. Mer,* **18,** 117–140.

STEEMANN NIELSEN, E. 1961. Chlorophyll concentration and rate of photosynthesis in *Chlorella vulgaris. Physiol. Plant.,* **14,** 868–876.

STEEMANN NIELSEN, E. 1962. On the maximum quantity of plankton chlorophyll per surface unit of a lake or the sea. *Int. Rev. ges. Hydrobiol.,* **47,** 333–338.

STEEMANN NIELSEN, E., and V. KR. HANSEN. 1959a. Light adaptation in marine phytoplankton populations and its interrelation with temperature. *Physiol. Plant.,* **12,** 353–370.

STEEMANN NIELSEN, E., and V. KR. HANSEN. 1959b. Measurements with the carbon-14 technique of the respiration rates in natural populations of phytoplankton. *Deep-Sea Res.,* **5,** 222–233.

STEEMANN NIELSEN, E., and V. KR. HANSEN. 1961. Influence of surface illumination on plankton photosynthesis in Danish waters (56°N) throughout the year. *Physiol. Plant.,* **14,** 595–613.

STEEMANN NIELSEN, E., and E. G. JØRGENSEN. 1962. The physiological background for using chlorophyll measurements in hydrobiology and a theory explaining daily variations in chlorophyll concentration. *Arch. Hydrobiol.,* **58,** 349–357.

STEEMANN NIELSEN, E., V. KR. HANSEN and E. G. JØRGENSEN. 1962. The adaptation to different light intensities in *Chlorella vulgaris* and the time dependence on transfer to a new light intensity. *Physiol. Plant.,* **15,** 505–517.

STEEMANN NIELSEN, E., and T. S. PARK. 1964. On the time course in adapting to low light intensities in marine phytoplankton. *J. Cons. Int. Explor. Mer,* **29,** 19–24.

STEEMANN NIELSEN, E., and E. G. JØRGENSEN. 1968. The adaptation of algae. I. General part. *Physiol. Plant.,* **21,** 401–413.

STEFÁNSSON, U., and F. A. RICHARDS. 1964. Distributions of dissolved oxygen, density and nutrients off the Washington and Oregon coasts. *Deep-Sea Res.,* **11,** 355–380.

STEPHENS, K. 1970. Automated measurement of dissolved nutrients. *Deep-Sea Res.,* **17,** 393–396.

STOMMEL, H. 1949. Trajectories of small bodies sinking slowly through convection cells. *J. Mar. Res.,* **8,** 24–29.

STRAIN, H. H. 1951. The pigments of algae. In *Manual of Phycology.* Chronica Botanica, Waltham, Mass., 243–262.

STRAIN, H. H. 1958. *Chloroplast Pigments and Chromatographic Analysis,* 32nd Priestley Lecture. Pennsylvania State University Press, pp. 180.

STRAIN, H. H. 1966. Fat-Soluble Chloroplast Pigments: Their Identification and Distribution in Various Australian Plants. In *Biochemistry of Chloroplasts.* Ed. T. W. Goodwin. Academic Press, New York, vol. 1, 387–406.

STRATHMANN, R. R. 1967. Estimating the organic carbon content of phytoplankton from cell volume or plasma volume. *Limnol. Oceanogr.,* **12,** 411–418.

STRICKLAND, J. D. H. 1958. Solar radiation penetrating the ocean. A review of requirements, data and methods of measurement, with particular reference to photosynthetic productivity. *J. Fish. Res. Bd. Canada*, **15**, 453–493.

STRICKLAND, J. D. H. 1960. Measuring the production of marine phytoplankton. *Fish. Res. Bd. Canada Bull.*, **122**, 172.

STRICKLAND, J. D. H. 1965. Production of organic matter in the primary stages of the marine food chain. In *Chemical Oceanography*, Eds. J. P. Riley and G. Skirrow. Academic Press, London, vol. 1, 477–610.

STRICKLAND, J. D. H. 1968. A comparison of profiles of nutrient and chlorophyll concentrations taken from discrete depths and by continuous recording. *Limnol. Oceanogr.*, **13**, 388–391.

STRICKLAND, J. D. H., and K. H. AUSTIN. 1960. On the forms, balance and cycle of phosphorus observed in the coastal and oceanic waters of the Northeastern Pacific. *J. Fish. Res. Bd. Canada*, **17**, 337–345.

STRICKLAND, J. D. H., and T. R. PARSONS. 1968. A practical handbook of seawater analysis. *Fish. Res. Bd. Canada, Bull.*, **167**, 311.

STRICKLAND, J. D. H., R. W. EPPLEY and B. ROJAS DE MENDIOLA. 1969. Phytoplankton populations, nutrients and photosynthesis in Peruvian coastal waters. *Bol. Inst. del Mar del Peru*, **2**, 1–45.

SUGIURA, Y. 1965. On the reserved nutrient matters. *Bull. Soc. Franco–japonaise d'Oceanographie*, **2**, 7–11.

SUSHCHENYA, L. M. 1962. Quantitative data on nutrition and energy balance of *Artemia salina* (L.). *Dokl. Akad. Nauk SSSR*, **143**, 1205–1207 (In Russian). English transl. *Dokl. (Proc.) Acad. Sci. USSR*, **143**, 329–330.

SUSHCHENYA, L. M. 1970. Food rations, metabolism and growth of crustaceans. In *Marine Food Chains*, Ed. J. H. Steele, Oliver and Boyd, Edinburgh, 127–141.

SUTCLIFFE, W. H. JR., E. R. BAYLOR and D. W. MENZEL. 1963. Sea surface chemistry and Langmuir circulation. *Deep-Sea Res.*, **10**, 233–243.

SUYAMA, M., K. NAKAJIMA and J. NONAKA. 1965. Studies on the protein and non-protein nitrogenous constituents of *Euphausia*. *Bull. Japanese Soc. Sci. Fish.*, **31**, 302–306 (Japanese with English summary).

SUZUKI, N., and K. KATO. 1953. Studies on suspended materials 'marine snow' in the sea. Part I. Sources of marine snow. *Bull. Fac. Fish., Hokkaido University*, **4**, 132–137.

SVERDRUP, H. U. 1953. On conditions for the vernal blooming of phytoplankton. *J. Cons. Explor. Mer*, **18**, 287–295.

SVERDRUP, H. U., M. W. JOHNSON and R. H. FLEMING. 1946. *The Oceans, their Physics, Chemistry and General Biology*. Prentice-Hall, New York, pp. 1087.

SYSOEVA, T. K., and A. A. DEGTEREVA. 1965. The relation between the feeding of cod larvae and pelagic fry and the distribution and abundance of their principal food organisms. *ICNAF Spec. Publ.*, **6**, 411–416.

SZEICZ, G. 1966. Field measurements of energy in the 0·4–0·7 micron range. In *Light as an ecological factor*. Eds. R. Bainbridge, G. C. Evans and O. Ractham. *Symp. Brist. Ecol. Soc.*, **6**, 41–51. Blackwell Scientific Publications, Oxford.

SZEKIELDA, K. 1967. Some remarks on the influence of hydrographic conditions on the concentration of particulate carbon in sea water. *IBP Symposium (Amsterdam)*. Eds. H. L. Golterman and R. S. Clymo. Noord-Hollandsche Vitgevers Maatschappi, Amsterdam, 314–322.

TAGUCHI, S., and K. NAKAJIMA. 1971. Plankton and seston in the sea surface of three inlets of Japan. *Bull. Plankton Soc. Japan*, **18**, 20–36.

TAKAHASHI, M., S. SHIMURA, Y. YAMAGUCHI and Y. FUJITA. 1971. Photoinhibition of phytoplankton photosynthesis as a function of exposure time. *J. Oceanogr. Soc. Japan*, **27**, 43–50.

TAKAHASHI, M., K. SATAKE and N. NAKAMOTO. 1972. Chlorophyll profile and photosynthetic activity in the north and equatorial Pacific Ocean. *J. Oceanogr. Soc. Japan*, **28**, 27–36.

TAKAHASHI, M., and T. R. PARSONS. 1972. The maximization of the standing stock and primary productivity of marine phytoplankton under natural conditions. *India J. Mar. Sci.*, **1**.

TALLING, J. F. 1957a. The phytoplankton population as a compound photosynthetic system. *New Phytol.*, **56**, 133–149.

TALLING, J. F. 1957b. Photosynthetic characteristics of some freshwater plankton diatoms in relation to underwater radiation. *New Phytol.*, **56**, 29–50.

TALLING, J. F. 1960. Comparative laboratory and field studies of photosynthesis by a marine planktonic diatom. *Limnol. Oceanogr.*, **5**, 62–77.

TAMIYA, H., E. HASE, K. SHIBATA, A. MITSUYA, T. IWAMURA, T. NIHEI and T. SASA. 1953. Kinetics of growth of *Chlorella* with special reference to its dependence on quantity of available light and on temperature. In *Algal culture: from laboratory to pilot plant*. Ed. J. S. Burlow. Carnegie Inst. Publ. No. 600, 204–232.

TANADA, T. 1951. The photosynthetic efficiency of carotenoid pigments in *Navicula minima. Amer. J. Bot.*, **38**, 276–283.

TEAL, J. M. 1962. Energy flow in the salt marsh ecosystem of Georgia. *Ecology*, **43**, 614–624.

TEIXEIRA, C., J. TUNDISI and J. SANTORO. 1967. Plankton studies in a mangrove environment. IV. Size fractionation of the phytoplankton. *Bolm Inst. Oceanogr. S. Paulo*, **16**, 39–42.

THOMAS, W. H. 1970a. On nitrogen deficiency in tropical Pacific oceanic phytoplankton: photosynthetic parameters in poor and rich water. *Limnol. Oceanogr.*, **15**, 380–385.

THOMAS, W. H. 1970b. Effect of ammonium and nitrate concentration on chlorophyll increases in natural tropical Pacific phytoplankton populations. *Limnol. Oceanogr.*, **15**, 386–394.

THOMAS, W. H., and R. W. OWEN, JR. 1971. Estimating phytoplankton production from ammonium and chlorophyll concentrations in nutrient-poor water of the eastern tropical Pacific Ocean. *Fish. Bull.*, **69**, 87–92.

TILTON, R. C. 1968. The distribution and characterization of marine sulfur bacteria. *Rev. Intern. Oceanogr. Med.*, **9**, 237–253.

TILTON, R. C., A. B. COBER and G. E. JONES. 1967. Marine thiobacilli. I. Isolation and distribution. *Can. J. Microbiol.*, **13**, 1521–1528.

TOMINAGA, H., and S. ICHIMURA. 1966. Ecological studies on the organic matter production in a mountain river ecosystem. *Bot. Mag., Tokyo*, **79**, 815–829.

TRANTER, D. J., and B. S. NEWELL. 1963. Enrichment experiments in the Indian Ocean. *Deep-Sea Res.*, **10**, 1–9.

TRAVERS, M. 1971. Diversité du microplancton du Golfe de Marseille en 1964. *Mar. Biol.*, **8**, 308–343.

TSYBAN, A. V. 1971. Marine bacterioneuston. *J. Oceanogr. Soc. Japan*, **27**, 56–66.

UDA, M. 1961. Fisheries oceanography in Japan, especially on the principles of fish distribution, concentration, dispersal and fluctuation. *Calif. Coop. Oceanic. Fish. Invest.*, **8**, 25–31.

VACCARO, R. F., and J. H. RYTHER. 1959. Marine phytoplankton and the distribution of nitrite in the sea. *J. Cons. Int. Explor. Mer*, **25**, 260–271.

VACCARO, R. F., and H. W. JANNASCH. 1967. Variations in uptake kinetics for glucose by natural populations in seawater. *Limnol. Oceanogr.*, **12**, 540–542.

VINOGRADOV, A. P. 1953. The Elementary Chemical Composition of Marine Organisms. Translation by Efron and Selton. *Memoir. Sears Found. Mar. Res.*, **2**, 130–146.

VINOGRADOV, M. E. 1955. Vertical migrations of zooplankton and their importance for the nutrition of abyssal pelagic fauna. *Trudy Inst. Okean.*, **13**, 71–76.

VINOGRADOVA, Z. A., and V. V. KOVAL'SKIY. 1962. Elemental composition of the Black Sea plankton. *Dokl. Akad. Nauk. SSSR*, **147**, 1458–1460.

VOGEL, K., and B. J. D. MEEUSE. 1968. Characterization of the reserve granules from the dinoflagellate *Thecadinium inclinatum* Balech. *J. Phycol.*, **4**, 317–318.

VOLLENWEIDER, R. A. 1965. Calculation models of photosynthesis—depth curves and some implications regarding day rate estimates in primary production measurements. In *Primary Productivity in Aquatic Environments*, Ed. C. R. Goldman. Mem. Ist. Ital., Idrobiol., *18* Suppl.: The University of California Press, Berkeley, 425–457.

VOLLENWEIDER, R. A. (Ed.). 1969. A manual on methods for measuring primary production in aquatic environments including a chapter on bacteria. *IBP Handbook No. 12*, F. A. Davis Company, Philadelphia, pp. 213.

VOLLENWEIDER, R. A., and A. NAUWERCK. 1961. Some observations on the ^{14}C-method for measuring primary production. *Verh. int. Ver. Limnol.*, **14**, 134–139.

VON ARX, W. S. 1962. *An introduction to physical oceanography*. Addison–Wesley, London, pp. 135.

WADA, E., and A. HATTORI. 1971. Nitrite metabolism in the euphotic layer of the central North Pacific Ocean. *Limnol. Oceanogr.*, **16**, 766–772.

WAILES, G. H. 1937. *Canadian Pacific fauna*. 1. Protozoa. 1a. Lobosa, 1b. Reticulosa, 1c. Heliozoa, 1d. Radiolaria, Biological Board of Canada, Toronto, pp. 14.

WAILES, G. H. 1939. *Canadian Pacific fauna*. 1. Protozoa. 1e. Mastigophora, Biological Board of Canada, Toronto, pp. 45.

WAILES, G. H. 1943. *Canadian Pacific fauna*. 1. Protozoa. 1f. Ciliata, 1g. Suctoria, Biological Board of Canada, Toronto, pp. 46.

WAKSMAN, S. A., C. L. CAREY and H. W. REUSZER. 1933. Marine bacteria and their role in the cycle of life in the sea. I. Decomposition of marine plant and animal residues by bacteria. *Biol. Bull.*, **65**, 57–79.

WALLEN, D. G., and G. H. GEEN. 1971a. Light quality in relation to growth, photosynthetic rates and carbon metabolism in two species of marine plankton algae. *Mar. Biol.*, **10**, 34–43.

WALLEN, D. G., and G. H. GEEN. 1971b. Light quality and concentration of proteins, RNA, DNA and photosynthetic pigments in two species of marine plankton algae. *Mar. Biol.*, **10**, 44–51.

WALLEN, D. G., and G. H. GEEN. 1971c. The nature of the photosynthate in natural phytoplankton populations in relation to light quality. *Mar. Biol.*, **10**, 157–168.

WALSH, J. J. 1971. Relative importance of habitat variable in predicting the distribution of phytoplankton at the ecotone of the antarctic upwelling ecosystem. *Ecol. Monographs*, **41**, 291–309.

WALSH, J. J. 1972. Implications of a systems approach to oceanography. *Science*, **176**, 969–975.

WALSH, J. J., and R. C. DUGDALE. 1971. A simulation model of the nitrogen flow in the peruvian upwelling system. *Inv. Resq.* **35**, 309–330.

WANGERSKY, P. J. 1965. The organic chemistry of sea water. *Amer. Scient.*, **53**, 358–374.

WANGERSKY, P. J., and D. C. GORDON. 1965. Particulate carbonate, organic carbon, and Mn^{++} in the open ocean. *Limnol. Oceanogr.*, **10**, 544–550.

WATSON, S. W. 1965. Characteristics of a marine nitrifying bacterium, *Nitrosocystis oceanus* sp. nov. *Limnol. Oceanogr. Suppl.*, **10**, (Redfield 75th Anniv. Vol.), R274–R289.

WATSON, S. W. 1971. Taxonomic considerations of the family Nitrobacteraceae Buchanan. *Int. J. System. Bacteriol.*, **21**, 254–270.

WATT, W. D., 1966. Release of dissolved organic material from the cells of phytoplankton populations. *Proc. Roy. Soc. (London)*, B, **164**, 521–551.

WEBB, K. L., and R. E. JOHANNES. 1967. Studies of the release of dissolved free amino acids by marine zooplankton. *Limnol. Oceanogr.*, **12**, 376–382.

WEBB, K. L., and R. E. JOHANNES. 1969. Do marine crustaceans release dissolved amino acids? *Comp. Biochem. Physiol.*, **29**, 875–878.

WEISS, P. 1969. The Living System: Determinism Stratified. In *Beyond Reductionism*, Eds. A. Koestler and J. Smythies. Hutchinson, London, 3–55.

WELCH, E. B., and G. W. ISAAC. 1967. Chlorophyll variation with tide and with plankton productivity in an estuary. *J. Water Poll. Cont. Fed.*, **39**, 360–366.

WESTLAKE, D. F. 1965. Some problems in the measurement of radiation under water: A review. *Photochemistry and Photobiology*, **4**, 849–868.

WHEELER, E. H. 1967. Copepod detritus in the deep sea. *Limnol. Oceanogr.*, **12**, 697–701.

WICKETT, W. P. 1967. Ekman transport and zooplankton concentration in the North Pacific Ocean. *J. Fish. Res. Bd.*, **24**, 581–594.

WICKSTEAD, J. H. 1962. Food and feeding in pelagic copepods. *Proc. Zool. Soc. London*, **139**, 545–555.

WIEBE, P. H. 1970. Small-scale spatial distribution in oceanic zooplankton. *Limnol. Oceanogr.*, **15**, 205–217.

WIEBE, P. H. 1971. A field investigation of the relationship between length of tow, size of net and sampling error. *J. Cons. Int. Explor. Mer*, **34**, 110–117.

WIEBE, P. H., and W. R. HOLLAND. 1968. Plankton patchiness: effects on repeated net tows. *Limnol. Oceanogr.*, **13**, 315–321.

WILHM, J. L. 1968. Use of biomass units in Shannon's Formula. *Ecology*, **49**, 153–156.

WILLIAMS, P. H. LEB. 1970. Heterotrophic utilization of dissolved organic compounds in the sea. I. Size distribution of population and relationship between respiration and incorporation of growth substrates. *J. mar. biol. Ass. U.K.*, **50**, 859–870.

WILLIAMS, P. M. 1965. Fatty acids derived from lipids of marine origin. *J. Fish. Res. Bd. Canada*, **22**, 1107–1122.

WILLIAMS, P. M. 1968. Organic and inorganic constituents of the Amazon River. *Nature*, **218**, 937–938.

WILLIAMS, P. M., and K. S. CHAN. 1966. Distribution and speciation of iron in natural waters: Transition from river water to a marine environment, British Columbia, Canada. *J. Fish. Res. Bd. Canada*, **23**, 575–593.

WILLIAMS, P. M., H. OESCHGER and P. KINNEY. 1969. Natural radiocarbon activity of the dissolved organic carbon in the North-east Pacific Ocean. *Nature*, **224**, 256–258.

WILLIAMS, P. M., and L. I. GORDON. 1970. Carbon-13: carbon-12 ratios in dissolved and particulate organic matter in the sea. *Deep-Sea Res.*, **17**, 19–27.

WILLIAMSON, M. H. 1961. A method for studying the relation of plankton variations to hydrography. *Bull. Mar. Ecol.*, **5**, 224–229.

WILSON, D. P. 1951. A biological difference between natural sea waters. *J. mar. biol. Ass. U.K.*, **30**, 1–20.

WINSOR, C. P., and G. L. CLARKE. 1940. A statistical study of variation in the catch of plankton nets. *J. Mar. Res.*, **3**, 1–34.

WOOD, E. J. F. 1953. Heterotrophic bacteria in marine environments of eastern Australia. *Aust. J. Mar. Freshw. Res.*, **4**, 160–200.

WOOD, E. J. F. 1958. The significance of marine microbiology. *Bact. Rev.*, **22**, 1–19.

WOOD, E. J. F. 1965. *Marine Microbial Ecology*. Reinhold, New York, pp. 243.

WOOD, H. G., and C. H. WERKMANN. 1935. The utilization of CO_2 by the propionic acid bacteria in the dissimulation of glycerol. *J. Bacteriol.*, **30**, 332.

WOOD, H. G., and C. H. WERKMAN. 1936. The utilization of CO_2 in the dissimulation of glycerol by the propionic acid bacteria. *Biochem. J.*, **30**, 48–53.

WOOD, H. G., and C. H. WERKMAN. 1940. The relationship of the bacterial utilization of CO_2 to succinic acid formation. *Biochem. J.*, **34**, 129–138.

WOOSTER, W. S., and J. L. REID, JR. 1963. Eastern boundary currents. In *The Seas*, Ed. M. N. Hill, Interscience Publishers, New York, vol. 2, 253–280.

WRIGHT, R. T. 1964. Dynamics of a phytoplankton community in an ice-covered lake. *Limnol. Oceanogr.*, **9**, 163–178.

WRIGHT, R. T., and J. E. HOBBIE. 1965. The uptake of organic solutes in lake water. *Limnol. Oceanogr.*, **10**, 22–28.

WRIGHT, R. T., and J. E. HOBBIE. 1966. Use of glucose and acetate by bacteria and algae in aquatic ecosystems. *Ecology*, **47**, 447–464.

WURSTER, C. F. 1968. DDT reduces photosynthesis by marine phytoplankton. *Science*, **159**, 1474–1475.

WURSTER, C. F., and D. B. WINGATE. 1968. DDT residues and declining reproduction in the Bermuda petrel. *Science*, **159**, 979–981.

YAMADA, M., and T. OTA. 1970. Studies on the lipid of plankton. IV. Unsaponifiable matter of lipid of *Calanus plumchrus*. *J. Japan Oil Chem. Soc.*, **19**, 377–382. (Fish. Res. Bd. Canada Trans. No. 1590.)

YENTSCH, C. S. 1965. Distribution of chlorophyll and phaeophytin in the open ocean. *Deep-Sea Res.*, **12**, 653–666.

YENTSCH, C. S., and J. H. RYTHER. 1957. Short-term variations in phytoplankton chlorophyll and their significance. *Limnol. Oceanogr.*, **2**, 140–142.

YENTSCH, C. S., and J. H. RYTHER. 1959. Absorption curves of acetone extracts of deep water particulate matter. *Deep-Sea Res.*, **6**, 72–74.

YENTSCH, C. S., and R. W. LEE. 1966. A study of photosynthetic light reactions, and a new interpretation of sun and shade phytoplankton. *J. Mar. Res.*, **24**, 319–337.

YOUNGBLOOD, W. W., M. BLUMER, R. L. GUILLARD and F. FIORE. 1971. Saturated and unsaturated hydrocarbons in marine benthic algae. *Mar. Biol.*, **8**, 190–201.

ZAITZEV, YU. P. 1961. Surface pelagic biocoenose of the Black Sea. *Zool. Zh.*, **40**, 818–825.

ZEITSCHEL, B. 1970. The quantity, composition and distribution of suspended particulate matter in the Gulf of California. *Mar. Biol.*, **7**, 305–318.

ZEUTHEN, E. 1970. Rate of living as related to body size in organisms. *Polskie Archiwum Hydrobiologii*, **17**, 21–30.

ZOBELL, C. E. 1946. *Marine Microbiology*. Chronica Botanica, Waltham, Mass., pp. 240.

ZOBELL, C. E. 1962. Geochemical aspects of the microbial modification of carbon compounds. In *Advances in Organic Geochemistry*, Pergamon Press, London, 1–18.

ZOBELL, C. E. 1968. Bacterial life in the deep sea. In Proceedings of the U.S.–Japan Seminar on Marine Microbiology. August 1966 in Tokyo. *Bull. Misaki Marine Biol. Kyoto Inst. Univ.*, No. 12, 77–96.

INDEX